TECHNOLOGICAL APPLICATIONS

ADVANCES IN LEARNING AND BEHAVIORAL DISABILITIES

Series Editors: Thomas E. Scruggs and
Margo A. Mastropieri

Recent volumes:

Volume 12: Edited by Thomas E. Scruggs and Margo A. Mastropieri

Volume 13: Edited by Thomas E. Scruggs and Margo A. Mastropieri

Volume 14: Educational Interventions – Edited by Thomas E. Scruggs and Margo A. Mastropieri

ADVANCES IN LEARNING AND BEHAVIORAL DISABILITIES
VOLUME 15

TECHNOLOGICAL APPLICATIONS

EDITED BY

THOMAS E. SCRUGGS
George Mason University, Fairfax, USA

MARGO A. MASTROPIERI
George Mason University, Fairfax, USA

United Kingdom – North America – Japan
India – Malaysia – China

Emerald Group Publishing Limited
Howard House, Wagon Lane, Bingley BD16 1WA, UK

British Library Cataloguing in Publication Data
A catalogue record for this book is available from the British Library

ISBN: 978-0-7623-0815-6
ISSN: 0735-004X (Series)

INVESTOR IN PEOPLE

CONTENTS

LIST OF CONTRIBUTORS

Jeffrey P. Bakken	Illinois State University, USA
Michael Behrmann	George Mason University, USA
Richard Boon	Fairfax County Public Schools, Fairfax, USA
Frederick J. Brigham	University of Virginia, USA
John Castellani	Johns Hopkins University, USA
Cesare Cornoldi	Università di Padova, Italy
Rossana De Beni	Università di Padova, Italy
Dave L. Edyburn	University of Wisconsin, Milwaukee, USA
Steve Forness	University of California, Los Angeles, USA
Jennifer J. Jakubecky	University of Virginia, USA
Tara Jeffs	Bowling Green University, USA
Kenneth Kavale	University of Iowa, USA
Kathy Klingerman	Indiana University, Indiana, USA
Margo A. Mastropieri	George Mason University, USA
Juanita Jo Matkins	University of Virginia, USA
Jackie McDonnouugh	University of Virginia, USA
Angelica Moè	Università di Padova, Italy
Lisa Mohler	Frankfort, Indiana, USA

James M. Royer University of Massachusetts, USA

Kenneth A. Rath University of Massachusetts, USA

Thomas E. Scruggs George Mason University, USA

Jennifer Shields University of Virginia, USA

Debra Sprague George Mason University, USA

Loel N. Tronsky University of Massachusetts, USA

Evangelia Zaimi University of Virginia, USA

PART I:

TECHNOLOGICAL APPLICATIONS

AUTOMATICITY TRAINING AS A READING INTERVENTION FOR ADOLESCENTS WITH ATTENTIONAL DISORDERS

James M. Royer, Kenneth A. Rath and Loel N. Tronsky

ABSTRACT

Twenty-three adolescents diagnosed with attentional disorders and reading issues were introduced to an intervention program designed to increase their reading ability through automaticity training on lists of words. For those nine students who completed the eight-week intervention program, substantial improvements were found on word reading, non-word reading, category matching and sentence completion tasks. The high rate of attrition (over 50%) is, however, indicative of the problems associated with attentional issues.

The relationship between attentional disorders in children and youth and academic difficulties are well documented. Hinshaw (1992a) reviewed the literature examining the relationship between academic problems and externalizing behaviors, which included ADD, ADHD, Oppositional Defiance Disorders and Conduct Disorders. His review particularly focused on academic deficiencies that could not be explained by low IQ and defined academic difficulty as performance sufficiently low as to qualify for special educational

Technological Applications, Volume 15, pages 3–16.

ISBN: 0-7623-0815-x

services. He reported that the overlap between externalizing behavior problems in general and academic difficulties was slightly under 20%. However, he noted that when the focus was only on students with attentional disorders, the degree of overlap between behavioral problems and academic difficulties increased to a considerable extent. In fact, the relationship between attentional disorders and academic difficulty was stronger than was the relationship between subaverage intelligence and academic difficulty. If the criterion for academic difficulties was loosened to include difficulties that did not meet learning disability standards, most students with attentional disorders would qualify as having academic problems (Hinshaw, 1992a). Indices of academic difficulty in this looser category would include poor grades, retention, suspension, and academic deficits.

The most common form of academic difficulty experienced by children with attentional disorders is reading delay, and there is some evidence that there is a chain of behaviors that leads from early language difficulties (primarily expressive rather than receptive), to attentional disorders to problems in reading acquisition (Silva, 1987; Stevenson, 1984). A causal chain in this linkage has not been established (Hinshaw, 1992a). It could be that there is a common agent that underlies the cognitive and behavioral symptoms. For example, in some children there appears to be a syndrome wherein early externalizing behaviors are accompanied by language difficulties, and the co-occurrence of these are strongly predictive of subsequent academic underachievement (Hinshaw, 1992b). However, since problems in externalizing behavior seem to increase as children enter the school years and begin to experience academic failure, it could also be that academic failure either encourages or exacerbates externalizing behaviors.

Hinshaw (1992b) has also commented on the developmental trajectory of the behavior of children with attentional disorders. His review shows a strong degree of comorbidity between attentional disorders and academic problems, and then as age advances, a tendency for youth with ADD or ADHD to develop oppositional defiance disorders and conduct disorders. Hinshaw (1992b) has also reviewed evidence showing that attentional disorders are resistant to interventions that are designed to alter patterns of externalizing behavior and/or academic difficulties. In particular, interventions focused only on the externalizing behavior or the academic difficulty show little evidence of effectiveness (Hinshaw, 1992b).

One probable reason that attentional disorders are so resistant to intervention is that the behavioral and academic difficulties exacerbate one another. Students with attentional disorders fail to engage in behaviors that produce academic success. The subsequent lack of academic success leads to even less inclination to engage in academic behaviors, which in turn tends to encourage the further occurrence of behaviors that are inconsistent with academic success.

The notion of reciprocal reinforcement of attentional difficulties and academic failure suggests that an effective treatment model should include elements that are targeted at both behavioral and academic problems. Hinshaw (1992b) makes this point quite strongly when he notes that, "children with comorbid achievement and behavior problems require intervention that combines the best of behavioral programming and educational instruction" (p. 899).

The study to follow reports the results of an academic intervention that was targeted at improving one aspect, the reading performance of students with attentional disorders. Some of these children and youth were also engaged in private psychotherapy that was attempting to lessen the occurrence of oppositional behaviors at home and school. Under ideal circumstances we would be able to report the effectiveness of both the behavioral and the academic intervention. Unfortunately, given the nature of the referral system our laboratory operates under we do not have information on the effectiveness of any behavioral interventions our participants were involved in. Thus, our research does not allow a true separation of outcomes associated with children experiencing both psychotherapy and academic intervention and those only experiencing academic interventions.

Approach to Reading Intervention

The automaticity training model we use in our intervention approach is based on the assumption that skilled reading necessarily involves the development of a number of low level skills that function automatically. In particular, the model is based on the idea that letter recognition, word recognition, semantic activation, and syntactic processing in the skilled reader all function automatically and with little conscious control. Modularity theory captures these assumptions (e.g. Stanovich, 1990). Modularity theory proposes that repeated reading practice transforms an activity controlled by conscious cognitive processes into one that automatically activates upon exposure to a particular stimulus event. These automatic processes are encapsulated in specialized processing modules that are arranged in serial order. Letter recognition can serve as an example. Early in reading development a student is told that a particular letter ("a") has a name and is often also told that the letter makes a particular sound that corresponds to a sound in spoken language. Initially, the letter name and sound is stored in declarative memory and the learner has to go through a conscious search strategy each time the letter is encountered. This search strategy is time consuming and error prone.

As practice identifying letters increases the learner begins to form a memory representation that binds together the visual orthographic stimulus, the name associated with the stimulus, and the speech sound that the letter makes. With continued practice the variety of stimuli that will activate this representation increases (generalizes to other fonts and to handwritten stimuli) and the process increasingly becomes data-driven. That is, the memory representation for the letter will activate even when the learner might be trying to consciously inhibit activation of the representation. The Stroop effect is a well-known example of exactly this process. As practice with reading increases, specialized modules develop that identify words, activate the meaning of words, and perform syntactic analysis of the text being read.

The development of fast and automatic performance of low level activities is essential to skilled reading because of the capacity limitations of working memory. Working memory is limited in the amount of information it can hold, and more importantly for the reading process, can only hold information for a short period of time without conscious rehearsal. Most items of information will decay from working memory in under 15 seconds if they are not refreshed via rehearsal or re-inspection. This time limit is important to reading comprehension where the meaning of a series of words must be accumulated in working memory before the meaning of the unit (a phrase or sentence) can be interpreted. If the reader has to consciously engage in an activity like identifying letters, words, the meaning of words, or syntactic analysis, it increases the likelihood that early processed words will decay from working memory before a meaningful unit can be interpreted.

As an example of the comprehension difficulties conscious processing produces, numerous studies show that students with specific reading disorders (dyslexia) are very slow at word identification (e.g. Royer & Sinatra, 1994). Problems with word identification subsequently result in difficulties with reading comprehension because early words decay before a meaningful unit can be accumulated and interpreted in working memory.

The automaticity training model we use assumes that many of the difficulties students are experiencing with reading comprehension will disappear if low level processes can be automated. Our strategy for making low-level processes automatic involves practicing the speeded processing of text materials that are carefully selected to match the skill level of the student.

THE AUTOMATICITY TRAINING MODEL

The Laboratory for the Assessment and Training of Academic Skills (LATAS) at the University of Massachusetts has developed an intervention model that involves identifying reading skills that are poorly developed and then providing focused practice that is designed to strengthen the skills. Poorly developed skills are identified using the Computer-based Academic Assessment System (CAAS). The CAAS system measures the speed and accuracy of performance of both reading and math activities (see Royer & Tronsky, 1998 for a description of intervention effectiveness with math disabled students). The reading battery measures the speed and accuracy of letter identification, word identification, concept activation, and semantic interpretation of sentences. We have two versions of our CAAS tasks. One version has been normed and we administer the norm version for the initial assessment, and then at four-week intervals in the intervention process. A second version of the CAAS system contains materials that are different than those in the norm version. This version, called the weekly version, is administered on the off weeks between administrations of the CAAS norm version. All of the tasks in the CAAS system contain materials that are randomly sampled for each test administration. For instance, the norm version of the CAAS word task contains 260 words, 40 of which are randomly sampled for presentation of any given test occasion. We use two versions of the CAAS system and the stimulus sampling procedure to assess transfer of training skills. That is, students are training with materials that are different than those in the CAAS system. This allows a clean evaluation of whether training activities are transferring to untrained materials.

After an assessment has been completed, the data is transformed into grade level normative performance, thereby allowing the identification of reading skills that are subaverage for a student's grade level. The lowest level reading skill is then targeted for intervention and if and when performance on that skill improves to the point where it is near or at grade level, the intervention focus moves to the next higher skill in the hierarchy of reading skill. So, for example, if the CAAS assessment indicated that a student was slow and/or inaccurate in letter identification relative to grade peers, the initial intervention would focus on improving the speed and accuracy of letter identification. When letter recognition reached a grade appropriate level, the intervention effort would shift to word identification, and so on.

The CAAS system provides profiles of performance that are distinctly different for different types of learning difficulties (Royer, 1997). For instance, learners with a specific reading disability have a profile that is different from the profile exhibited by a garden variety poor reader, and the profile of both of these in turn are quite different from the profile of a reader with a subaverage IQ. Royer (1997) also shows that the reading profile for a learner with attentional problems is also different from the profile of a reader with a specific reading disability.

Training Rapid and Accurate Word Identification

Most of the learners we have worked with who have reading problems exhibit slow and/or inaccurate word identification abilities. Automaticity training is an intervention approach designed to make word identification fast and accurate. It begins with the selection of practice words at the appropriate level of difficulty. The CAAS assessment provides evidence useful for making this decision. The practice words are typically sets of 160 words divided into forty words per page. Five times per week the student practices naming the words as fast as possible while being as accurate as possible. The student names the words on one page as fast as possible while maintaining accuracy, and the time per page (in seconds) is recorded. Each of the remaining pages is practiced and the average time per page computed and recorded on a graph. Practice on the set of words continues until the time levels off with minimal error rates, and the student is then given a new set of words. After several sets of words at a given level of difficulty are mastered, the difficulty level of the words is increased and practice continues.

Practice at rapid word naming typically occurs either at home or at school, but students return to LATAS once a week where they repeat the weekly CAAS assessments. Every four weeks the students repeat the CAAS norm assessment. The materials in the CAAS norm assessment are different than those the student is practicing on a daily basis. This allows an evaluation of whether improvement on the practice activities is associated with specific practice, or with general improvement in reading skill.

With this general introduction to the assessment and intervention procedures, we will now turn to a description of a study that uses these procedures to improve the reading skills of students with attentional and academic problems.

METHODOLOGY

Participants

Table 1 provides descriptive information for all of the participants discussed in this article. The age is given in years and months and, along with the grade,

Table 1. Participants Who Completed a CAAS Assessment and Eight Weeks of Intervention.

Name	Age	Sex	Grade	Referral Source	Nature of Diagnosis and Interventions
A. C.	12y 0m	F	7	Word of Mouth	Information unavailable Was taking medication
H. A.	10y 3m	F	4	Clinical Psychologist A	Diagnosed by A and receiving family therapy for behavioral issues
H. C.	10y 9m	M	5	Word of Mouth	Initially diagnosed by school Currently taking medication Also diagnosed as dyslexic
K. J.	12y 8m	M	7	Neuropsychologist A	Diagnosed as borderline ADHD by neuropsychologist A Also diagnosed as dyslexic
M. R.	11y 3m	M	4	Word of Mouth	Diagnosed by a neuropsychologist A. Some indication of bipolar disorder Currently taking medication for ADHD
R. N.	13y 4m	M	8	Clinical Psychologist A	Diagnosed by A and working on behavioral issues Received medication in the past but not currently
R. J.	8y 8m	M	2	Clinical Psychologist A	Diagnosed by A and receiving family therapy for behavioral issues Also diagnosed as dyslexic
S. J.	12y 6m	M	6	Clinical Psychologist A	Diagnosed by A and receiving family therapy for behavioral issues
C. A.	10y 10m	M	5	Clinical Psychologist A	Diagnosed by a physician Receiving medication since Grade 1

represents the participant's status at the time of the initial assessment. The "referral source" column describes the individual (if recorded) who suggested the LATAS program to the participant's family. The "nature of diagnosis and interventions" column describes, in brief, how the individual was diagnosed with ADD/ADHD and what other diagnoses and treatments (if any) were being received at the time of the initial assessment. The participants who completed the eight-week intervention included seven males and two females, with an average age of 11 years 3 months.

Computer-based Assessments

Each of the participants in the study initially completed computer-based assessments of reading performance using the CAAS tasks below.

Word Naming
Words of varying difficulty appear on the screen and the student says the name of the word into the microphone. The CAAS elementary word task contains 240 words, 40 of which are randomly sampled (10 from 4 categories varying in difficulty) to conduct a CAAS assessment.

Pseudoword Naming
Pronounceable non-words (produced by altering words in the word task) appear on the screen and the student pronounces the nonword into the microphone. This task is identical to the word task except for the stimuli.

Concept Activation
The student is told that pairs of words will appear on the screen that either belong or do not belong to a named category. When the words appear, the student says "yes" (they belong to the same category) or "no." Again it is the case that assessments are conducted by sampling from a large pool of items.

Semantic Processing of Sentences
A sentence appears on the screen that has a blank in it and a word appears above the blank and below the blank. The words vary in semantic appropriateness and the student says the name of the word into the microphone. Assessments are conducted by sampling from a large pool of items.

Reading Intervention

Our reading intervention, which was briefly described earlier, consisted of providing the student with four pages of forty words at an appropriate level of difficulty and with a graph that could be used to record the average reading time per page (in seconds) on a daily basis. Each student practiced rapid word naming at least five days per week. A practice session consisted of having the student look over the words and asking for assistance identifying words that were not immediately recognizable. When the students were confident that they knew all of the words they indicated they were ready to be timed. They then read the words on the page as fast as possible while trying to maintain accuracy. If they did not know a word they were instructed to guess. A helper (parent, teacher, fellow student) timed the word reading in seconds and made note of words that were missed. After the page had been completed the helper returned to words that were missed and provided the correct name for the word. Words that were missed repeatedly over several practice sessions were placed on a separate list and practiced both before and after a timing session.

The timing activity was repeated for each of the pages and upon completion of the four pages of words the student calculated average time per page and plotted it on the graph. The order in which words were named on the page (left to right, up or down) was varied for each practice session to prevent responding to the words from memory.

Average time per page always dropped within the first three days of practice and practice on a set of words continued until time leveled off and remained constant for 3–5 days. Most students level off between 30 and 40 seconds per page (slightly under a second per word). When the students reached the level performance criterion for mastery they were given a new set of 160 words and practice continued. It should be mentioned that watching the line on the graph drop was the most effective reinforcer in the intervention process.

CAAS Re-assessments

Once a week students would return to our laboratory and complete another CAAS assessment. As mentioned earlier, we used two sets of materials to complete the assessments: our regular weekly tasks and the once-every-four-weeks norm tasks. The materials in the CAAS assessments were different than the practice materials. The data to be reported in the results section come from the norm assessments, which were administered on the initial visit and on weeks 4 and 8 during the intervention process.

RESULTS

Table 2 shows the speed and accuracy of initial performance on each of the CAAS tasks. The table also reports grade level percentile performance for each of the tasks. The percentiles are based on a combined speed/accuracy index. As can be seen in the table, the reading performance of the participants prior to intervention was generally well below average for their grade level. The exception is CA who was above average on word naming and sentence comprehension performance, but below average on nonword naming and

Table 2. Accuracy, Response Time, and Grade Level Percentile on Each of the CAAS Reading Tasks.

Name		Word	Non-Word	Category	Sentence
A.C.					
	% Correct	94	76	100	100
	RT (sec)	1.09	1.26	3.07	6.76
	Grd %tile	**14**	**7**	**2**	**1**
H.A.					
	% Correct	89	83	95	91
	RT (sec)	0.99	2.59	2.31	4.82
	Grd %tile	**19**	**5**	**35**	**9**
H.C.					
	% Correct	92	94	94	85
	RT (sec)	1.04	2.23	5.06	6.05
	Grd %tile	**9**	**7**	**1**	**4**
K.J.					
	% Correct	97	78	100	92
	RT (sec)	2.91	3.19	2.26	5.86
	Grd %tile	**1**	**1**	**8**	**1**
M.R.					
	% Correct	95	89	89	100
	RT (sec)	1.10	1.26	3.1	6.0
	Grd %tile	**47**	**63**	**12**	**5**
R.N.					
	% Correct	86	97	100	86
	RT (sec)	0.98	1.15	1.87	3.61
	Grd %tile	**1**	**39**	**11**	**12**
R.J.					
	% Correct	55	61	66	77
	RT (sec)	8.08	13.9	13.28	18.7
	Grd %tile	**1**	**1**	**1**	**1**
S.J.					
	% Correct	100	58	100	100
	RT (sec)	0.86	1.3	1.7	2.9
	Grd %tile	**12**	**1**	**43**	**50**
C.A.					
	% Correct	97	80	90	100
	RT (sec)	0.55	0.96	1.4	2.2
	Grd %tile	**75**	**17**	**15**	**71**

category performance. The reader should also note that the three students who were diagnosed with both an attentional disorder and dyslexia (HC, KJ and RJ) were particularly poor at the reading tasks.

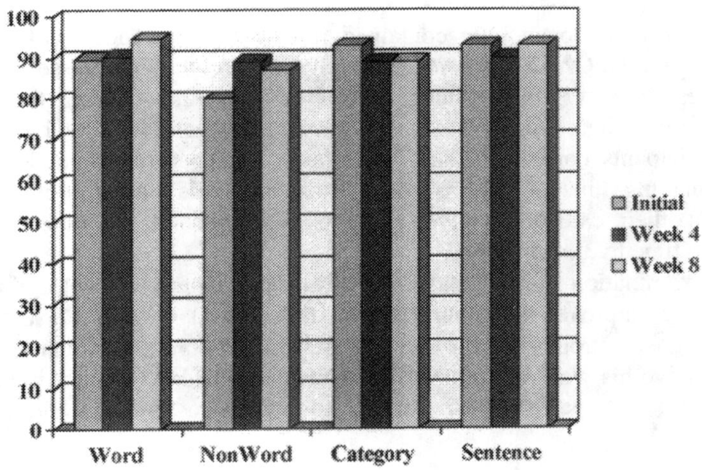

Fig. 1. Accuracy Changes During Intervention.

The results of the reading intervention can be seen in Figs 1 and 2. Figure 1 shows accuracy performance prior to the intervention and after four and eight weeks of intervention. Figure 2 shows the impact of the intervention on average

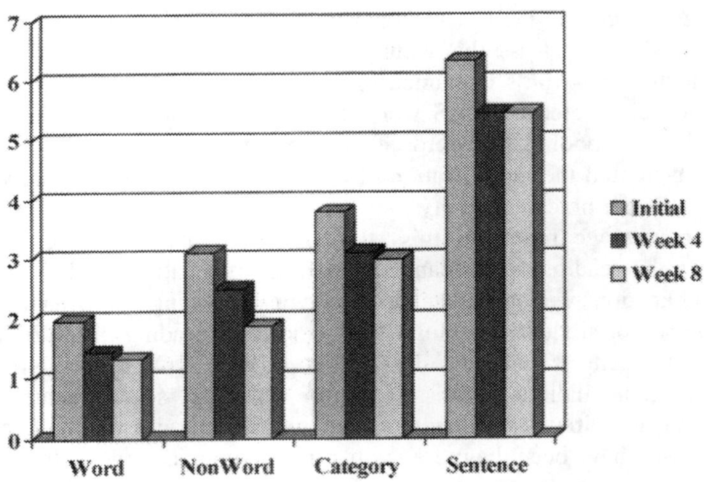

Fig. 2. Response Time Changes During Intervention.

time to respond to an item contained in a task. The reader will note that accuracy on the CAAS tasks was generally high on the initial assessment and that there was very little change in response accuracy during the 8 week intervention. There was, however, substantial changes in the speed with which the participants completed the CAAS tasks. The participants reduced their word naming time by 0.63 seconds, their nonword naming time by 1.23 seconds, their category analysis time by 0.80 seconds and their sentence analysis time by 0.88 seconds.

An examination of the data of individual participants indicated that every participant improved their performance (got faster) on every task with the following exceptions. AC did not improve his nonword performance, MR did not improve his word and nonword performance, and RJ did not improve his sentence speed performance, though his sentence accuracy did improve slightly.

DISCUSSION

The results of the intervention indicated that practicing the rapid naming of words generally improved the participant's ability to identify words, to identify non-words, to activate the meaning of words, and to process and understand sentences. It should be emphasized that each of these improvements reflected transfer from the practice activity rather than improvement associated with direct practice. Even the increase in the speed with which the participants recognized words on the CAAS word task reflected transfer of an improving cognitive skill. That is, the words the participants were practicing were different than the words contained in the CAAS norm task. Moreover, each participant completed a CAAS word task that contained 40 words randomly sampled from a pool of 240 words. Thus, the transfer from the practice activity directly benefited the identification of a large number of words that were not contained in the practice activity.

The results we report in this study indicate that automaticity training improves the reading performance of students with attentional disorders. In other publications we report that the same type of training improves the reading performance of students diagnosed with a specific reading disability (Cisero, Royer, Marchant & Jackson, 1997; Royer, 1997). Our procedure is not, however, an unqualified success. Over the course of several years we have completed an initial assessment on, and offered services to, a total of 23 students who have been diagnosed with an attentional disorder. The 9 students we describe in this study are the participants who accepted the offer of services and who stayed with the program for at least 8 weeks. The remaining 14 students declined the offer of services after they were described, or discontinued the practice activities after a few weeks. It should be emphasized that in virtually all of the 14 cases the decision to discontinue services was made by the participant rather than by a parent. In all of the cases the parents were eager to either initially try the intervention or to continue with the program after it had been started.

Our less than 50% success rate at maintaining the participation of students with attentional disorders in automaticity training is far less than we have experienced when working with students diagnosed with dyslexia or dyscalculia. There are probably two reasons for this. First, the attentional disorder students who are referred to our laboratory are typically older than their counterparts with specific academic difficulties. This means that they have had long periods of experiencing academic difficulties and therefore probably more inclination to seek sources of self-esteem in activities that are non-academic in nature. The second reason that many of our attentional disorder students chose not to participate in our intervention is that they are often oppositional to any suggestion made by an adult authority figure (including their parents).

We believe that the techniques described in this article could be a valuable addition to the repertoire of procedures designed to improve the reading performance of students with attentional disorders. However, it is also obvious that these techniques will only work if the students are willing to use them, and getting to use them will most likely involve the use of behavioral interventions designed to enhance academic motivation and the willingness to engage in academically productive activities.

REFERENCES

Cisero, C. A., Royer, J. M., Marchant, H. G., & Jackson, S. J. (1997). Can the Computer-based Academic Assessment System (CAAS) be used to diagnose reading disability in college students? *Journal of Educational Psychology, 89*, 599–620.

Hinshaw, S. P. (1992a). Externalizing behavior problems and academic underachievement in childhood and adolescence: Causal relationships and underlying mechanisms. *Psychological Bulletin, 111*, 127–155.

Hinshaw, S. P. (1992b). Academic underachievement, attention deficits, and aggression: Comorbidity and implications for intervention. *Journal of Consulting and Clinical Psychology, 60*, 893–903.

Royer, J. M. (1997). A cognitive perspective on the assessment, diagnosis and remediation of reading skills. In: G. Phye (Ed.), *Handbook of Academic Learning* (pp. 199–234). San Diego, CA: Academic Press.

Royer, J. M., & Sinatra, G. M. (1994). A cognitive theoretical approach to reading diagnostics. *Educational Psychology Review, 6*, 81–113.

Royer, J. M., & Tronsky, L. N. (1998). Addition practice with math disabled students improves subtraction and multiplication performance. *Advances in Learniing and Behavioral Disabilities*, *12*, 185–217.

Silva, P. A. (1987). Epidemilogy, longitudinal course and some associated factors (an update). In: I. W. Yule & M. Rutter (Eds), *Language Development and Disorders* (pp. 1–15). Oxford, England: MacKeith.

Stanovich, K. E. (1990). Concepts in developmental theories of reading skill: Cognitive resources, automaticity and modularity. *Developmental Review*, *10*, 72–100.

Stevenson, J. (1984). Predictive values of speech and language screening. *Developmental Medicine and Child Neurology*, *26*, 528–538.

ZONING IN ON PHYSICS: CREATING VIRTUAL REALITY ENVIRONMENTS TO AID STUDENTS WITH LEARNING DISABILITIES

Debra Sprague and Michael Behrmann

INTRODUCTION

Providing access to abstract concepts in science for students with learning disabilities has been educationally difficult. Students with learning disabilities typically exhibit problems with reading fluency, text comprehension skills, vocabulary learning, and abstract reasoning from text presentations (Scruggs & Mastropieri, 1993). In fact, these learners read science text at only about half the fluency rate as students without disabilities (Parmar, Deluca & Janczak, 1994).

By providing experiential, three-dimensional representations for those concepts that cannot be readily understood in conventional instructional formats, immersive virtual environments can increase learning disabled students' access to the regular secondary level physics curriculum. Access to this curriculum and the acquisition of science literacy can lead to improved standards-based learning, more technical and scientific educational experiences and career choices, and high motivation for individuals with learning disabilities.

This chapter explores the benefits and barriers of virtual reality and the use of virtual reality with students identified as having a learning disability or

Technological Applications, Volume 15, pages 17–38.

ISBN: 0-7623-0815-x

emotional disorder. The chapter will describe the design of *Zoning In On Physics*, a virtual reality environment that addresses foundations of physics instruction for students with learning disabilities.

DEFINITION OF VIRTUAL REALITY

Imagine students being able to walk on Mars without a spacesuit, talk to Socrates or Galileo, explore the relationship between mass, force, and friction in a world without gravity, or become a test particle and experiment with electrical fields. Such opportunities could make history and literature come alive for students or enable them to understand scientific concepts in a way they never could before. Such opportunities are possible with the help of an emerging technology called virtual reality.

The term "virtual reality" has been used to describe many different types of computer generated, or "virtual" interactions, from text-based interactions in chat rooms to video arcade games played through a two-dimensional screen. For the purpose of this chapter we are defining virtual reality (VR) as a three-dimensional, computer-generated synthetic environment that gives the user a sense of being immersed in a real world. This immersion is based on visual, audio, and haptic (touch) feedback. Instead of using standard keyboards and screens, people wear head-mounted displays, data gloves, and headphones. A computer controls what they hear, see, and feel. They, in turn, control the computer by manipulating the objects they see, feel and hear, while interacting within the synthetic environment (Aukstakalnis & Blatner, 1992).

In order for students to interact and move around within the virtual environment, they must use specialized equipment. One of the first pieces of equipment a user needs is a head-mounted display. Such a device looks like a helmet with a giant pair of goggles mounted on it. Head-mounted displays allow the user to see the virtual environment. They also contain tracking devices that tell the computer where the user is looking.

Another piece of equipment needed for full immersion is headphones, which allow the user to hear sounds that take place in the virtual environment. Headphones can be built directly into the head-mounted displays and do not need to be a separate device. It is also necessary to have equipment that allows the user to manipulate the virtual environment. This can be accomplished in multiple ways. One of the most common forms of navigation is the data glove. This is a glove with sensors that fits over the user's hand. When the user moves his real hand, the glove picks up the movement and sends an electronic signal to the computer. This signal is translated into the motions of a virtual hand.

Another way to accomplish the same task is to use a 3Ball. This device is simply a No. 3 billiard ball hollowed out to house a magnetic triaxial sensor. A 3Ball also includes a small button that can be pressed to select objects and specify commands. The user can move a cursor or object around in VR by holding the 3Ball in his hand and moving it about (Aukstakalnis & Blatner, 1992).

Researchers are also experimenting with the development of equipment that will produce haptic feedback to the user of the virtual environment. One such equipment is a belt that contains vibrating pagers similar to the ones used in restaurants. Developed by researchers at George Mason University, the belt is used to help students understand the relationship between source charges and electrical fields. When the user (who becomes a source charge) approaches a strong electric field, the pagers begin to vibrate. When the user backs away from the field, the vibrations stop (Norton & Sprague, 2001). There are also haptic trackballs where resistance to movement increases with distance and speed.

For all of this equipment to work together and allow the user to feel immersed in the virtual world, a high-end computer is needed. Such a computer must have a faster processor and more memory than what is found in most of the standard computers used in schools today.

CHARACTERISTICS OF VIRTUAL REALITY

Virtual reality has been used to train pilots to fly and respond to emergencies. It has been used to help doctors understand the human body and develop skills needed to perform surgery (Aukstakalnis & Blatner, 1992). VR has been used to help people overcome phobias, to train children with disabilities to operate a wheelchair (Inman, Peaks, Loge, Chen & Ferrington, 1994), and to help engineers and architects design airplanes and houses. If VR is appropriate for training people in specific skill areas, it seems that VR would also be a good tool to educate people in abstract concepts. Various research conducted on VR's potential in education has revealed positive results, showing it to be a potentially positive and engaging learning experience (Sprague, 1996).

VR offers a number of characteristics that may prove beneficial for students with learning disabilities and emotional disorders. Many of these characteristics have been used effectively in a variety of teaching strategies. However, VR provides unique and powerful ways of employing them (Powers & Darrow, 1996).

Multisensory

Virtual reality, like the real world it emulates, has the ability to present information through multiple senses. As stated earlier, VR allows the user to see, hear, and feel the world around them. For students with visual or auditory processing deficits, providing information through multiple senses is a necessity. Such multisensory cues can direct a student's attention to important relationships, such as the relationship between source charges and electrical fields or the relationship between speed and friction. Using multisensory cues (sight, sound, and haptic) can enhance the quality of the student's learning and interaction experiences (Dede, Salzman, Loftin & Sprague, 1999).

Experiential Learning Environment

Virtual reality can be characterized as an experiential learning environment. Students actively interact with the environment. Learning-by-doing combined with reflective inquiry can induce learning; through experience, students are able to extend and modify their understanding of science concepts (mental models) based on discontinuities between expected and actual behaviors of phenomena (Dede, Salzman, Loftin & Ash, 1997).

VR allows the user to ask "what if . . ." questions. It allows the user to test out his hypothesis and immediately see the outcome. Students can compare initial conditions and end results. They can perform the same experiment over and over until they fully comprehend the results.

Abstract to Concrete

In order to understand the world and how it functions, students must comprehend abstract concepts far removed from their everyday experiences. Mastery of abstract science concepts often requires learners to incorporate invisible factors that represent intangible forces (diSessa, 1983). However, real-life experiences often distort or contradict the concepts students need to understand. For example, the presence of friction makes objects in motion seem to slow and stop "on their own," undercutting the face validity of Newton's First Law, "*In the absence of an unbalanced force, an object at rest remains at rest, and an object already in motion remains in motion with constant speed in a straight line path*". As a result, most learners have difficulty understanding these scientific concepts (Reif & Larkin, 1991).

Virtual reality allows users to explore alternative realities because, unlike the real world, the laws of nature or the laws of physics do not bind it. Conditions such as gravity or friction can be turned off so students can see the outcomes when these conditions are no longer an issue. By turning off these conditions, students are able to identify their misconceptions about the ways in which the world functions. Once identified, students begin to alter their misconceptions. This leads to students developing new mental models that demonstrate their ability to comprehend these abstract scientific concepts.

Control of Extraneous Stimuli

Students with learning difficulties often find it impossible to manage extraneous stimuli. They are not always clear on what is important and what can be ignored. VR allows for the creation of worlds in which extraneous stimuli can be controlled. Using a scaffolding process, students may initiate learning in a highly simplified environment. As proficiency increases, the environment can become increasingly complex until it resembles the real world. Such a progression of environments in which stimuli is increased would be difficult to achieve in a traditional classroom (Middleton, 1992).

Through the use of the head-mounted display and headphones, it is also possible to eliminate distractions from the real world. Such distractions often interfere with learning for students with learning and/or emotional disabilities. Without such distractions, the user is forced to attend to the stimuli presented in the VR environment. This could increase the amount of time on task and the amount of learning that will occur.

Frames-of-Reference

One of the unique capabilities of virtual reality is being able to look at an environment from multiple viewpoints. This is referred to as frames-of-reference. In virtual environments, students can become part of a phenomenon and experience it directly. Alternatively, they can step back from the phenomenon to allow a global view of what is happening (Salzman, Dede, Loftin & Chen, in press). Experiencing the world directly is referred to as egocentric while stepping back is referred to as exocentric (McCormick, 1995; Wickens & Baker, 1995). Enabling students to experience phenomena from multiple perspectives or frames of references appears to facilitate the learning process (Dede, Salzman, Loftin & Ash, 1997).

By using frames-of-reference (FORs) in virtual reality, learners are provided with experiences that they would otherwise have to imagine. "One frame-of-reference may make salient information that learners might not notice in another frame-of-reference. Further, multiple frames-of-reference might help students to fill in gaps in their knowledge and to become more flexible in their thinking" (Dede et al., 1997, pp. 405–406).

BARRIERS TO VR

Despite these characteristics that make virtual reality beneficial for learning, there are barriers that make it difficult for VR to be used widespread in education. These barriers include: cost of the equipment, memory and graphic issues, simulator sickness, lack of well designed VR environments for educational use, and a lack of research conducted on VR use for students with learning disabilities or emotional disorders.

Cost of the Equipment

Because of the cost of the computer and equipment needed, full-immersive VR is not currently available in most schools. The development of high-end virtual environments with sophisticated scientific models requires computer equipment costing over $100,000 (Castellani, 1999). VR systems are expensive and far exceed the budgets of most school districts. However, the prices of VR systems are coming down due to advances in the entertainment industry. New developments in the field are producing low-cost alternatives that can be just as effective as full-immersive VR.

Recently, new software development tools have been designed for Windows NT operating systems running on high-end Pentium computers. Developers are now able to create immersive virtual environments that, through the use of a player (software developed by the company that allows users to interact with the environment created, but not make any changes to it), can run on Windows 95 and Windows 98 computers.

Instead of using head-mounted displays, a low-cost alternative is a pair of glasses called "CrystalEyes." CrystalEyes, developed by StereoGraphics, looks like a pair of sunglasses. They are outfitted with shutters that close every 1/100th of a second, producing high-definition 3-dimensional (3-D) images. CrystalEyes allow students to view a virtual world created on a two-dimensional screen in 3-D.

Instead of using a data glove or 3Ball, students use a standard computer mouse or an inexpensive haptic mouse to interact with the virtual world. As a result, schools are now able to take advantage of the virtual reality capabilities to enhance learning without cost being an impediment.

Memory and Graphic Issues

In order to produce VR environments that look and sound real, high-end computers are required. These computers must have faster processors and more memory than what is found in the standard computers used in schools today. The trade-offs for inexpensive systems are low-resolution graphics, problems with modeling complex scenes, and a lag between user motion and the response time of the system (Holloway, Fuchs & Robinette, 1991).

Currently, graphics and sound elements are constrained by the power and expense of the system. Virtual worlds are cartoons compared to the animated computer graphics we see on television and in the movies. Virtual reality is not able to provide realistic people to interact with as portrayed in the holodecks of *Star Trek: The Next Generation.*

Slow response time of the computer system is another limitation of VR. The refresh rate of a screen could be as infrequent as two or three times per second. Such a slow speed not only affects convenience, but also the user's health. At a display rate of less than 60 frames per second, there is a lag time between when the user's inner ear registers movement and when the system changes the view. Such a lag time can make the user suffer from a type of motion sickness called simulation sickness (Industry visualizes real uses for virtual worlds, 1993).

Simulation Sickness

Simulation sickness can result in feelings of disorientation, headaches, eyestrain, nausea, and vomiting. Simulation sickness can be brought on by conflicting cues provided by your senses and changes in the virtual environment. Like the lag time between the refresh rate and the inner ear's registration of movement, simulation sickness can also occur if the system is too real. In this situation, your eyes are telling you that you are moving but your body is saying you are not. This feeling is similar to the one you might feel if you were watching a movie filmed from a helicopter or airplane on a very large screen, such as an IMAX theater (Aukstakalnis & Blatner, 1992). Although researchers are aware of the correlation between conflicting cues and the occurrence of simulation sickness, they have not been able to identify the specific factors that cause this condition.

Simulation sickness is a concern when using VR, but it does not happen to everyone. In fact, it occurs in less than 2% of the population. Being aware of the risks involved and taking some precautions can help people not to experience simulation sickness. People susceptible to other forms of motion sickness may want to avoid VR. In addition, people who have an ear infection, cold, or epilepsy should avoid VR. As a precaution, women who are pregnant should also refrain from participating in VR. Having users take frequent breaks and checking on the status of their health before, during, and after the VR experience can help to reduce simulation sickness (Norton & Sprague, 2001).

Lack of VR Environments for Educational Use

Currently, there are few fully immersive VR environments designed for educational use in K-12. Limited access to systems for research and development due to high costs reduce the number and range of potential users (Powers & Darrow, 1996). However, with the lowering cost of the systems, there are a few projects that have been or are currently being developed for K-12 use.

The Virtual Gorilla Exhibit Project

The Virtual Gorilla Project is exploring the use of virtual reality to help people learn experientially what would otherwise be difficult to learn. Researchers at the Georgia Institute of Technology, in conjunction with Zoo Atlanta, are creating the project. Based upon actual data provided by Zoo Atlanta (gorilla behavior data as well as terrain data and building blueprints), the researchers are modeling a gorilla exhibit where the user can explore areas that are normally off limits (i.e. night quarters) to visitors of the Zoo (Allison & Hodges, 1999).

Through the use of this virtual exhibit, children can "be a gorilla" and experience first hand what it is like to be a member of a gorilla family group. They can walk around and interact with the virtual gorillas. In addition, they can test behaviors and elicit appropriate responses from the gorillas based upon the children's behavior and assumed role in the hierarchy of the family structure. To learn more about the Virtual Gorilla Exhibit Project visit their website at http://www.cc.gatech.edu/grads/a/Don.Allison/gorilla/.

ScienceSpace

Researchers at George Mason University and the University of Houston, through the sponsorship of the National Science Foundation and NASA, have teamed up to create three VR Worlds (NewtonWorld, MaxwellWorld, and PaulingWorld) in a project called ScienceSpace. The purpose of ScienceSpace is to access VR's potential and limitations as a learning tool. The virtual worlds are designed to help students master challenging scientific concepts. NewtonWorld allows students to explore Newton's Laws of Motion as well as kinetic energy and linear momentum. MaxwellWorld allows students to examine the nature of electrostatic forces and fields and helps them understand the concept of electrical flux. PaulingWorld enables students to explore the structure of both small and large molecules from different viewpoints. Research conducted on the virtual worlds in ScienceSpace show that the use of multisensory cues (visual, auditory, and haptic) could engage learners and direct their attention to important behaviors and relationships (Dede, Salzman & Loftin, 1996).

Although the majority of the research completed to date on ScienceSpace has been conducted in a laboratory setting, these VR worlds were placed in schools during the 1999–2000 academic year. Researchers were interested in seeing how teachers used these systems in their classrooms. NewtonWorld was placed in a 4th grade and a 6th grade classroom, while MaxwellWorld was placed in three high school classes. Teachers used VR as part of in-class activities, lab activities, and after school activities. Ongoing research is looking at the use of VR as an assessment tool. To learn more about ScienceSpace visit their website at http://www.virtual.gmu.edu.

Little to No Research Conducted for Students with Disabilities

The remaining barrier to the use of VR centers on the lack of research conducted with students who have learning disabilities or emotional disorders. The research community has largely ignored this population. The majority of the research conducted with special needs students has focused on those with physical disabilities or mental retardation (Roblyer & Cass, 1999).

One research study has looked at the use of VR with autistic children. In this study researchers used two children with mild to moderate autism and placed them in a VR environment designed to teach preliminary skills needed to cross a street. Immersion in the system was limited to five minutes to avoid simulation sickness. The children soon became adept in the environment and were able to take virtual walks and locate objects. The researchers reported that the children "responded similarly to three different street scenes, but more study needs to be done to determine if they were generalizing across different surroundings" (Strickland, Marcus, Mesibov & Hogan, 1996, p. 658).

Another project involving the design of virtual reality environments for students with moderate to severe learning disabilities is the Virtual Reality Applications Research Team (VIRART). This group of researchers and programmers are associated with the University of Nottingham. They have created four VR environments: House, City, Supermarket, and Skiing. The Virtual House consists of a kitchen, dining area, living room, and bedroom. The kitchen is fully interactive with cupboard doors that open, a stove that can be turned on, and a working sink. Here students learn how to make a cup of tea. In the living room, students will find a working television and telephone while in the bedroom is a wardrobe containing a range of clothes that can be chosen depending on the weather conditions (Brown, Standen & Cobb, 1998).

In the Virtual City students can choose from multiple frames of reference, including that of driving a car around the city, a pedestrian learning where and when it is safe to cross a street, or a person in a wheelchair investigating where ramp access should be. Virtual people occupy the city and walk around pre-set routes. In the Virtual Supermarket, students can enter a medium size supermarket and choose from a range of 60 products. The students can select the goods to buy by using a mouse or touch screen. When students proceed to the checkout, a coinage system appears so students can select the correct amount of money needed to pay for the goods. In Virtual Skiing, students are faced with a course complete with gates that they are to circumnavigate. The students ski down the slope, accelerating as they go, trying to pass through the center of each gate. At the bottom of the ski slope skiers are automatically attached to the ski lift, which takes them back to the top of the slope. Students can choose from three different speeds, with the faster being chosen as the skill and coordination of the student increases (Brown, Standen & Cobb, 1998).

VIRART conducted research to see if students with severe learning disabilities could use virtual environments to learn skills that will transfer to the real world. Twenty-three students between the ages of 15 and 19 participated in the study. The results showed that the students in the experimental group who used the Virtual Supermarket performed faster and more accurately in the real supermarket than did students in the control group who were not exposed to the Virtual Supermarket (Brown, Standen & Cobb, 1998).

There has also been some research conducted on adults with learning disabilities and the development of their vocational skills. One such study involved the creation of a virtual kitchen to assess the feasibility of using VR in vocational training of people with learning disabilities. There were four food preparation and cooking tasks and twelve potential hazards (i.e. a toaster with a frayed flex, a puddle on the floor) distributed around the virtual kitchen. Real task performance before and after the virtual kitchen training, real kitchen training, workbook training, and no training were compared. Participants benefited more from the virtual training than from the workbook training in the food preparation tasks but not in the hazard recognition tasks. The results showed that for people with learning disabilities, active interactions with a virtual environment can produce better learning than passive observations and that this learning can transfer to real world situations (Rose and Brooks, 2000).

PROJECT DEVISE: A CASE STUDY IN VR DEVELOPMENT IN SPECIAL EDUCATION

In the fall of 1999, a group of faculty from the special education and instructional technology departments at George Mason University (GMU) received a grant from the United States Department of Education Steppingstones to Technology Grant Program (*Designing Environments for Virtual Immersive Science Education*, PR #: H327A990049A). The purpose of the grant is to apply the principles of virtual reality-based instruction to meet the needs of high school students with mild disabilities. In reviewing the need for such a project, the faculty found that in general, textbooks expect the child with language and literacy difficulties to "go too fast, use too much vocabulary, and require too much reading and writing" (Brownell & Thomas, 1998, p. 121). In fact, textbook-oriented learning is the predominant approach used in science classrooms, particularly at the secondary level. Science textbooks have been found to be particularly difficult to read (Chiang-Soong & Yager, 1993), and can contain more new vocabulary than found in foreign language courses (Yager, 1983). The documented outcomes on science achievement for students with learning disabilities, compelled to try to learn from textbooks, have not been positive:

- Parmar, Deluca and Janczak (1994) found that students with learning disabilities read science text at only about half the fluency rate as students without disabilities.
- Carlisle and Andrews (1993) reported that students with learning disabilities performed significantly lower than their peers on a science curriculum-based assessment. These students also rated themselves, and were rated by their teachers, more negatively.

In response to such problems, researchers have suggested that activities-oriented (or "hands-on)" methods and materials were likely to interact more positively with the characteristics of learning disabilities (Mastropieri & Scruggs, 1994; Patton, 1993, 1995; Parmar, Deluca & Janczak, 1994; Scruggs & Mastropieri, 1994). Activities-oriented materials typically place fewer demands on language and literacy abilities and verbal memory, and provide relevant activities as learning experiences.

Purpose of Project DEVISE

As stated earlier, using VR, students with learning disabilities can *directly experience* the scientific concepts being studied, without the necessity of drawing abstractions from text they cannot read. Specifically, VR systems can address the special needs of students with learning disabilities in several ways:

- VR presentations can demonstrate complex concepts experientially, without reliance upon text and verbal lecture presentations.
- VR presentations employ not only the visual stimulation of models, but also auditory and haptic stimulation to allow the learner to truly experience the phenomena being studied (Salzman, Dede & Loftin, 1996).
- VR presentations are interactive, allowing the learner to proceed at an optimal pace, focus on more difficult aspects, and repeat experiences as many times as necessary to complete understanding.

The instructional rationale for developing virtual learning environments such as DEVISE is to provide students with learning disabilities a learning tool that: (1) provides multi-sensory experiences related to science concepts; (2) provides smaller units of instruction that address foundational concepts; (3) enables students to synthesize information based on multi-sensory input; (4) reduces the need for verbal expression and relies on other individual strength areas; (5) integrates current, age appropriate, real life science experiences into student centered learning activities; and (6) applies the principles of universal design to provide access to all students with disabilities (Castellani, 1999).

With these goals in mind, three organizations at GMU, the Graduate School of Education's (GSE) Helen A. Kellar Institute for Human disAbilities (KIHd), the GSE Instructional Technology Program, and the School of Information Technology, collaborated to provide faculty and student resources with content expertise in disabilities and science education; instructional design and development expertise; and technical computer programming and hardware development knowledge and skill. This collaboration has resulted in the development of a VR based software package, *Zoning In On Physics*, to teach the concepts inherent in Newton's Laws of Motion to high school age youth with mild learning disabilities.

Analysis and Design of Zoning In On Physics

The design process for instructional software directed toward individuals with disabilities is not a simple accomplishment. Adding the dimension of a new three-dimensional technology has made this project even more challenging. The Steppingstones to Technology grant program was initiated by U.S.DOE to enable development of "validated" effective uses of technology for children and youth with disabilities. An iterative approach to design and development in this project was indicated. Each of four student/faculty design teams (one per semester over two years) engaged in the analysis and design process for *Zoning In On Physics* at different levels. The first design team developed the broad needs analysis and overview of the project, conducting an initial literature review and brainstorming on the overall design of the project. Their major challenge as designers was to become aware of and document the special services offered to students with learning disabilities as well as the instructional accommodations afforded by the physics teacher. They had to gain an understanding of the makeup and structure of a mainstream physics class, gain an understanding of the *Active Physics* curriculum (the curriculum used in the local school district) and the overall instructional approach of the physics teacher with respect to Newton's laws, and identify common misconceptions the students have related to Newtonian physics.

The second team continued with the analysis of learners and the environments in which the DEVISE software, now entitled *Zoning In On Physics*, would be used. They reached the following conclusions about learner needs, which were important to the design of the project:

(1) Learners with mild disabilities are best engaged by activity-based instruction rather than formal lecture.
(2) Optimal instruction time prior to activity is brief (0–10 minutes).
(3) Directions should be delivered in small chunks, with the ability to be repeated as needed.
(4) Continual scaffolding is necessary for learners to make meaningful connections to activity content. (Scaffolding can be provided by teachers or by graphic representation and positive feedback in the software in varying degrees).
(5) Multi-sensory and multiple perspectives in delivery enhances instructional effectiveness.
(6) Positive verbal reinforcement and messages increase learner confidence.
(7) Learners need ability to progress at their own pace.
(8) Learners need ability to manipulate variables in activity in order to increase understanding and engagement.
(9) The learners are likely to bring similar misconceptions about Newtonian mechanics to the instructional setting as are seen in the general population of learners. Instruction will need to effectively address this issue.

This team developed the initial prototype design for *Zoning In On Physics*, setting up four "zones" of instruction. They also developed a storyboarding system that was able to communicate the multi-dimensional and multi-sensory intent of the prototype to the programmers. The prototype software, although

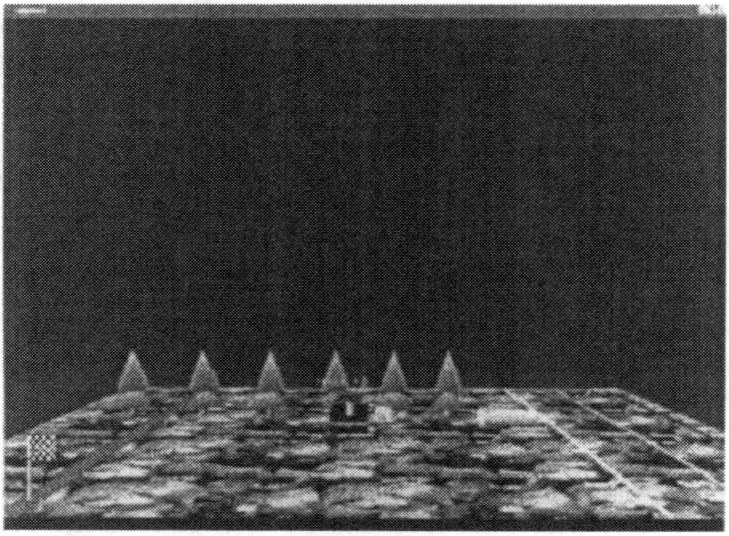

Fig. 1. Early Design of *Zoning In On Physics.*

utilizing simple two-dimensional graphics and simple sound files, provided a sound basis for further development by teams three and four (Fig. 1).

The third and fourth teams then engaged in a process of testing and refining the software. They conducted a formative evaluation with a small group of tenth and eleventh grade students with and without disabilities, obtained feedback from physics teachers and special educators, and worked with other subject matter experts to refine the software prototype. The third team refined the graphics and added graphs to provide visual feedback to students. They also refined the teacher guides and developed a website for product dissemination and feedback. The fourth team conducted more formal testing of the software with a larger group of high school students with mild disabilities and developed several intermediate instructional steps (chunks) within the four instructional zones, including the ability to manipulate the forces and direction vectors in a more realistic manner. They also added the 3-D visual and auditory elements that were more characteristic of VR and made the instructional software more appealing and authentic. This team also developed a plan to include kinesthetic feedback through a haptic mouse (to provide feedback related to force and direction) as well as for including surround-sound audio (to enhance auditory feedback) in future development. A future Steppingstones project is planned to test the efficacy of *Zoning In On Physics* in a wide array of school environments and with a large population of students with and without mild disabilities.

Zoning In On Physics

Zoning In On Physics consists of four modules or zones designed to focus Newton's Laws of Motion. Each zone builds on concepts taught in previous zones. *Zoning In On Physics* runs on Windows NT, 95, and 98 platforms. It can be viewed in 2-dimensions, or with the addition of CrystalEyes (http://www.qualixdirect.com/products/3dacc/ce3.htm), in 3-dimensions. Currently, the developers are looking at additional peripherals, such as stereo-surround sound and haptic mouse, to capitalize on the full potential of VR.

Several features are consistent within each of the four zones. There is a shuttle that students can move by applying force. To apply force, students need to click on the force button at the bottom of the screen (Fig. 2). Force can be applied in 100, 200, or 400 intervals. Students can also adjust the amount of friction the shuttle will encounter. Friction is determined by choosing the ground texture. Students can choose between outer space (no friction), ice (low friction), grass (medium friction), and bricks (high friction). Finally, students can choose the frame of reference for viewing the shuttle. Students may choose an exocentric (looking at the shuttle from the outside) or egocentric (looking at the shuttle from within) point of view. In the exocentric frame of reference, students may view the shuttle from the side, front, back, or bird's eye (looking down from the sky). In the egocentric frame of reference, students have a forward and backward view.

As students experiment with force and friction, they have the option to allow the program to graph their results. The graphs will record the speed of the shuttle and the distance traveled (Fig. 3). Graphs can be turned on or off with a click of a button. It is possible to show the graphs for the last five experiments, thereby allowing students to try different force and friction combinations and compare the results.

Zone 1

Zone 1 focuses on Newton's First Law of Motion, often called the Law of Inertia. This law states "an object at rest remains at rest until acted upon by a force or an object in motion continues moving in a straight line at constant velocity until acted upon by a force." In Zone 1, students are able to experiment with the relationship between force and friction. This zone has a battering ram, which students use to administer force on the shuttle. As students increase the

Fig. 2. Controls for *Zoning In On Physics*.

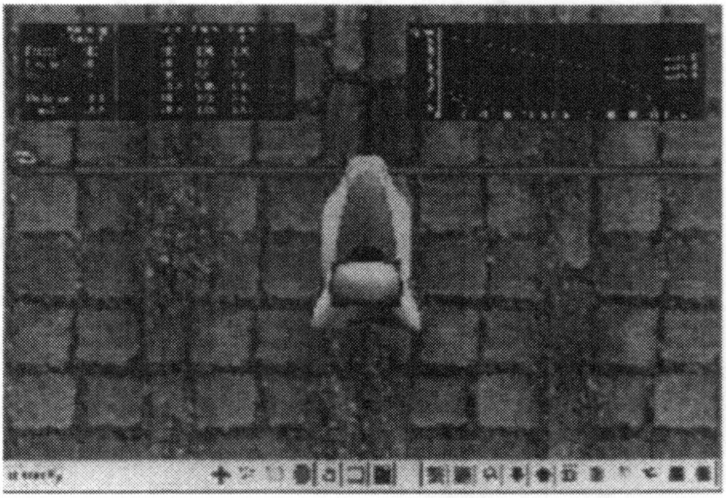

Fig. 3. Zoning In On Physics with Graphs Activated.

force, the battering ram moves back. When students press the button to release the battering ram it moves forward and hits the shuttle propelling it to move forward (much in the same way a pinball is propelled forward by the plunger hitting it). The shuttle will continue to move forward until it either stops due to friction or the students stop it. Students are able to choose how much friction there is by clicking on the various ground textures. In the experiments, students are asked to keep force constant while changing friction. They are asked to record the distant the shuttle traveled.

Zone 2

Zone 2 builds on the learning objective of Zone 1 and adds the concept of mass. Zone 2 focuses on Newton's Second Law of Motion which states "acceleration of an object is directly proportional to the net force acting on the object, and inversely proportional to its mass." This zone operates the same way as Zone 1 and includes the battering ram. However, a new feature is added. Here students can increase the mass of the shuttle by adding "mass balls." Each mass ball equals one unit of mass (Fig. 4). In these experiments, students are asked to keep force and friction constant while increasing and decreasing the amount of mass.

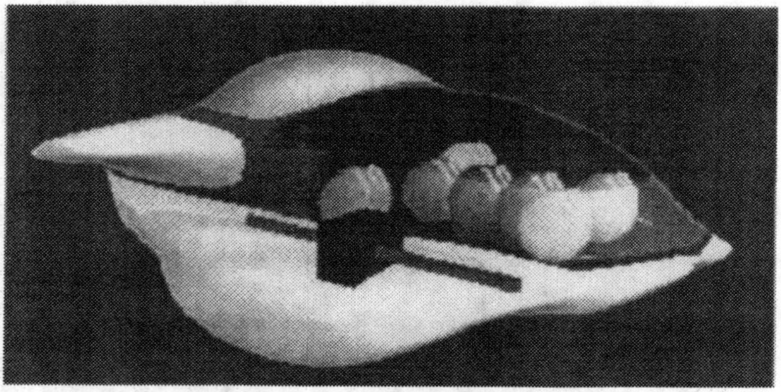

Fig. 4. Shuttle with Mass Balls.

Zone 3

Zone 3 focuses on Newton's Third Law of Motion which states "whenever one object exerts a force on a second object, the second object exerts an equal and opposite force on the first." Put another way, "for every action there is an opposite and equal reaction." The design on this zone is slightly different from the previous two zones. In this zone there is no longer a battering ram. Instead the shuttle has been equipped with a thrust device called a launcher. The launcher swivels around and shoots out puffs of gas. This propels the shuttle in the opposite direction. The launcher is able to swivel 360 degrees so the shuttle can go in any direction. Instead of an environment that allows the shuttle to move in one direction only, students now have X and Y axes labeled North, South, East, and West (Fig. 5). The amount of force is determined by how many puffs of gas shoots out of the launcher. In this zone, students experiment with direction, force, friction, and mass.

Zone 4

Zone 4 builds on the learning objectives of the previous zones and introduces a new concept – centrifugal force. Here students are presented with the same configuration as Zone 3. However, instead of the shuttle moving in a straight line, the shuttle now moves in a giant circle (Fig. 6). Students are given the variables for mass and friction and are asked to determine the amount of force needed to propel the shuttle completely around the circle. Once students have this concept, the parts of the walls of the circle can be removed. This allows students to experiment with decreasing and increasing force in order to keep the shuttle within the circle.

Fig. 5. Shuttle Going East in Zone 3.

Fig. 6. Shuttle Moving in the Circle in Zone 4 with Graphs Activated.

Zoning In On Physics is available free of charge from the project website at http://it.gse.gmu.edu/projects/ziop. Here teachers will find the four modules, the player that will allow the modules to run on various Window Platforms, lesson plan ideas, and student worksheets. As new modules are developed they will be added to the website.

CONCLUSION

This chapter explored the use of virtual reality to enhance the learning of students with learning disabilities and emotional disorders. VR has the ability to present information through multiple senses, is an experiential learning environment, enables the exploration of abstract concepts, provides multiple frames of references, and allows for the control of extraneous stimuli. All of this can result in a deeper understanding of physics for students who have difficulty reading the science textbook.

Although the cost of immersive VR systems is still beyond the budget of most school districts, new developments in the field are producing low-cost alternatives that can be just as effective as full-immersive VR. These alternatives are allowing developers to produce 3-dimensional VR worlds that are able to run on standard Window NT, 95, and 98 computers. One such program is *Zoning In On Physics,* which students can use to explore and develop an understanding of Newton's Laws of Motion.

ACKNOWLEDGMENTS

The authors wish to thank the following people for their assistance in the design and development of *Zoning In On Physics*: Dr. Christopher Dede, Dr. Debra Sprague, Dr. Michael M. Behrmann, Dr. Jim Chen, Dr. John Castellani, Ying Zhu, Jennie Schaff, Xusheng Wang, Joann Wray, Laura Dines, Shane Gallagher, Keysha Gamor, Colleen Kehoe, Eric Ritland, Lisa Saavedra, Linda Garner, Tomeka Gibbs, Dana Grabiner, Moira McGuiniss, Monica Villigran, Marc Zolar, Lori Cole, Rosemary Craft, Ruihua Dong, Jerry Fernandez, Sharon Greenspan, Komar Khan, Alesha Pulsinelli, Cynthia Roubie, Pam Tiffany, Colby Chambers, Nechele Hill, Jennifer Korjus, Margie Joyce, Joe McCahill, Kathryn Spence, Dr. Margo Mastropieri, Dr. Thomas Scruggs, Ben Allen, Heather Bousman, Dr. Donna Sterling, and Dr. Bowen Loftin.

REFERENCES

Allison, D., & Hodges, L. F. (1999). Virtual Gorilla Project, *AERA Symposium on Cross Project Research in Virtual Environments*, AERA 1999 Annual Meeting, Montreal, Canada.

Aukstakalnis, S., & Blatner, D. (1992). *Silicon mirage: The art and science of virtual reality.* California: Peachpit Press, Inc.

Brown, D. J., Standen, P. J., & Cobb, S. V. (1998). Virtual environments special needs and evaluative methods. In: G. Riva, B. K. Wiederhold & E. Molinari (Eds), *Virtual Environments in Clinical Psychology and Neuroscience.* Amsterdam, Netherlands: Ios Press.

Brownell, M. T., & Thomas, C. W. (1998). An interview with Margo Mastropieri: Quality science instruction for students with disabilities. *Intervention in School and Clinic, 34,* 118–122.

Carlisle, J. F., & Andrews, E. (1993). Monitoring learning-disabled students in mainstream science classes. *Annals of Dyslexia, 43,* 217–237.

Castellani, J. (1999). Project DEVISE, George Mason University Steppingstones Grant Proposal, http://www.virtual.gmu.edu/EDIT792/proposal.html

Chiang-Soong, B., & Yager, R. E. (1993). Readability levels of the science textbooks most used in secondary schools. *School Science & Mathematics, 93,* 24–27.

Dede, C., Salzman, M., & Loftin, B. (1996). *ScienceSpace: Virtual realities for learning complex and abstract scientific concepts.* Proceedings of IEEE Virtual Reality Annual International Symposium.

Dede, C., Salzman, M., Loftin, B., & Ash, K. (1997). Using virtual reality technology to convey abstract scientific concepts. In: M. J. Jacobson & R. B. Kozma (Eds), *Learning the Sciences of the 21st Century: Research, Design, and Implementing Advanced Technology Learning Environments.* Hillsdale, NJ: Lawrence Erlbaum.

Dede, C., Salzman, M., Loftin, R. B., & Sprague, D. (1999). Multisensory Immersion as a Modeling Environment for Learning Complex Scientific Concepts. In: N. Roberts, W. Fuerzeig & B. Hunter (Eds), *Computer Modeling and Simulation in Science Education.*

diSessa, A. (1983). Phenomenology and the evolution of intuition. In: D. Gentner & A. Stevens (Eds), *Mental Models* (pp. 15–33). Hillsdale, NJ: Lawrence Earlbaum Associates, Publishers.

Holloway, R., Fuchs, J., & Robinette, W. (1991). *Virtual-worlds research at the University of North Carolina at Chapel Hill.* Paper presented at the Conference on Computer Graphics, London, England.

Industry visualizes real uses for virtual worlds (1993). *Machine Design, 65*(15), 12–14.

Inman, D. P., Peaks, J., Loge, K., Chen, V., & Ferrington, G. (1994). Virtual Reality Training Program for Motorized Wheelchair Operation. *Proceedings from Technology and Persons with Disabilities Conference.* (March 16–19, 1994). Los Angeles. California State University, Northridge Center on Disabilities.

McCormick, E. P. (1995). *Virtual reality features of frames of reference and display dimensionality with stereopsis: Their effects on scientific visualization.* Unpublished master's thesis, University of Illinois at Urbana-Champaign, Urbana, Illinois.

Mastropieri, M. A., & Scruggs, T. E. (1994). Text-based vs. activities-oriented science curriculum: Implications for students with disabilities. *Remedial and Special Education, 15,* 72–85.

Middleton, T. (1992). Matching the Virtual Reality Solution to the Special Need. Paper presented at the *California State University at Northridge Technology and Persons with Special Needs Conference,* Los Angeles, CA.

Norton, P., & Sprague, D. (2001). *Technology for Teaching.* Needham Heights, MA: Allyn and Bacon, Inc.

Parmar, R. S., Deluca, C. B., & Janczak, T. M. (1994). Investigations into the relationship between science and language abilities of students with mild disabilities. *Remedial and Special Education (Rase), 15*(2), 117–126.

Patton, J. R. (1993). Individualizing for science and social studies. In: J. Wood (Ed.), *Mainstreaming: A Practical Approach for Teachers* (2nd ed., pp. 366–413). Columbus, OH: Merrill.

Patton, J. R. (1995). Teaching science to students with special needs. *Teaching Exceptional Children, 27*(4), 4–6.

Powers, D. A., & Darrow, M. (1996). Special Education and Virtual Reality: Challenges and Possibilities. *Journal of Research on Computing in Education, 27*(1).

Reif, F., & Larkin, J. (1991). Cognition in scientific and everyday domains: Comparison and learning implications. *Journal of Research in Science Teaching, 28*, 743–760.

Rose, D., & Brooks, B. (2000). The use of vitural reality in occupational training of people with learning disabilities. http://homepages.uel.ac.uk/E.A.Attree/mencap.html

Roybler, M. D., & Cass, M. (1999). Still more potential than performance: Virtual reality research in special education. *Learning and Leading With Technology, 26*(8), 50–53.

Salzman, M., Dede, C., Loftin, R. B., & Chen, J. (In Press). A model for understanding how virtual reality aids complex conceptual learning. *Presence: Teleoperators and Virtual Environments.*

Salzman, M. C., Dede, C., & Loftin, B. (1996). ScienceSpace: Virtual realities for learning complex and abstract scientific concepts. In *Proceedings of IEEE Virtual Reality Annual International Symposium* (pp. 246–253). New York: IEEE Press.

Scruggs, T. E., & Mastropieri, M. A. (1993). Current approaches to science education: Implications for mainstream instruction of students with disabilities. *Remedial and Special Education, 14*(1), 15–24.

Scruggs, T. E., & Mastropieri, M. A. (1994). The construction of scientific knowledge by students with mild disabilities. *Journal of Special Education, 28*, 307–321.

Sprague, D. (1996). Virtual reality and precollege education: Where are we today? *Learning and Leading With Technology, 23*(8), 10–12.

Strickland, D., Marcus, L. Mesibov, G., & Hogan, K. (1996). Brief report: Two case studies using virtual reality as a learning tool for autistic children. *Journal of Autism and Developmental Disorders, 26*(6), 651–659.

Wickens C. D., & Baker, P. (1995). Cognitive issues in virtual reality. In: W. Barfield & T. Furness (Eds), *Virtual Environments and Advanced Interface Design* (pp. 515–541). New York: Oxford University Press.

Yager, R. E. (1983). The importance of terminology in teaching K-12 science. *Journal of Research in Science Teaching, 20*, 577–578.

THE EYES MAY HAVE IT: RECONSIDERING EYE-MOVEMENT RESEARCH IN HUMAN COGNITION

Frederick J. Brigham, Evangelia Zaimi,
Juanita Jo Matkins, Jennifer Shields,
Jackie McDonnouugh and Jennifer J. Jakubecy

ABSTRACT

The study of eye movements relative to perception and cognition has been an area of interest from ancient times. The present paper will trace this research activity from its origins in the Ancient Greek and Medieval times through the 19th century and the technological progress of the 20th century. We will identify turning points in the history of eye movement and provide information on recently developed eye-tracking equipment. The discussion will include eye movements and visual perception, identifying the basic terminology and phenomena considered relative to eye movements (fixations, saccades, and regressions), visual perceptual and attentional spans. Emphasis will be placed on eye movements and comprehension during text- and picture-viewing and how judgements can be made, based on eye movements, about how people try to integrate content from text and pictures. We will then describe the fundamental conclusions made 15 to 20 years ago regarding the eye movements of individuals with reading difficulties as well as recently emerging evidence that calls for a reconsideration of some of those conclusions. Finally, we

Technological Applications, Volume 15, pages 39–59.

ISBN: 0-7623-0815-x

will describe our current efforts at educational applications of eye-movement research involving applications to content displays on the World Wide Web.

The study of eye movements has a long and sometimes inglorious past. The Greeks believed that the eyes, in their never-ceasing movement, were the mirror of the soul (Pavlidis, 1991), a notion that was preserved through time. In situations where humans have a variety of intersensory information available to them, they usually rely on the visual sense above all others even if the visual information contradicts the other senses or common experience (Stein & Meredith, 1993). The tendency to rely on visual information over other senses is the basis for all illusions performed by magicians. The following section outlines the development of eye-movement research from the Middle Ages to the beginning of the 21st century.

A BRIEF HISTORY OF EYE-MOVEMENT RESEARCH

Medieval consideration of human thought favored vision above all other senses because it was believed to express the state of the psyche and the mind and was a medium of knowledge about other human beings and of one's inner life (Heller, 1988). Most of the work done in this period was completed by Arabian scientists. Discoveries attributed to these scientists include the observations that central visual fields are perceived more clearly than the peripheral areas, and that recognition of familiar words can occur even if some of the characters can not be seen clearly. Like many scientific endeavors active at the beginning of the middle ages, the study of eye gaze was virtually dormant by the end of the period. With the development of the camera obscura in the 17th century, however, researchers were able to glimpse different aspects of human visual perception and a renewed interest in visual perception began to grow (Stafford, 1994).

The Revival of Eye-Movement Research in the 19th Century

Many of the efforts in the 19th century were aimed at developing a more accurate determination of eye motion characteristics. Examinations conducted during this time period demonstrated that head movements to one side result in compensatory rotations of the eyes in the opposite direction, vertical movements can be executed faster than horizontal ones, and that monocular (single-eye) movements can be carried out faster than binocular (both-eye) movements (Heller, 1988). By 1850, researchers using letter recognition techniques had begun to estimate the span of visual acuity for a single glimpse of a stimulus. Little attention was paid to links between eye movements and cognition during this era.

By the late 1800s, researchers shifted their attention to the psychology of visual perception. Until this time, the common belief was that each letter was presented in turn to the fovea (the area of clear and focused vision on the retina) and the eyes moved in a smooth pattern from letter to letter. However, closer examinations of the eye movements in readers failed to detect sweeping patterns. Instead, the eyes were observed to move in leaps during reading (Stark, Giveen & Terdiman, 1991). One of the first mechanical devices for recording eye-movements was developed during the late 19th century. This device collected information by placing a blunt needle on the upper eyelid. The needle responded to ocular movements in a manner similar to a phonograph needle responding to the grooves of a phonograph record. The mechanical movements of the eye create vibrations in the needle which are transmitted to the researcher's ear by means of an amplifying membrane and a rubber tube. A short sound indicated a leap within a line, and a longer sound a return leap between lines. Data produced by this device suggested that readers divide lines into subsections of 10 to 12 characters that are read during rhythmical intervals (fixations) and estimated the duration of these intervals as 250 to 400 ms. The transition from one section to the next is accomplished by energetic jerks during which reading cannot occur (Heller, 1988). Recent research has confirmed these early findings.

Technological Progress in Eye-Tracking Technology

The beginning of the 20th century brought the first photographic recording techniques to eye-movement research. In the early electronic systems, a beam of light was used to illuminate the cornea and the reflected light, focused by means of a lens system, was recorded on a moveable photographic plate (Heller, 1988). The development of sensitive and reliable eye-tracking systems is one of the major reasons for the relative success of eye-movement research (Rayner, 1992). Such systems provide researchers with substantial control over the stimulus environments and link data collection devices with high-speed computers for information processing. It is somewhat ironic that a brief session of eye-movement recording can produce a large amount of data when the individual being studied is rarely aware to the number of eye movements that are made. Schiller (1998) estimated that the number of saccades made by an individual in a single day could easily exceed 170,000. Without computer-based data, analysis the amount of information generated in a single eye-movement observation session would be very difficult to summarize and interpret.

Types of Eye-tracking Systems

Eye-tracking systems collect information about eye movements in one of three basic methods:

(1) Measuring the reflection of light reflected by the eye.
(2) Electro-oculography (measuring the electric potential of the skin around the eyes).
(3) Applying a special contact lens that facilitates tracking of its position.

Selection of an appropriate eye-tracking technology requires a balance between the need for precision and ecological factors such as intrusiveness of the technology. Highly precise eye-tracking systems are often the most intrusive for the user. For example, some systems require the user to immobilize his or her head by holding a bite bar in the mouth (e.g. Fletcehr, 1991) or placing the chin on a chin rest (e.g. Biscaldi, Gezek & Stuhr, 1998). Immobilizing the head ensures that visual orienting will be carried out only with eye movements, thereby increasing the accuracy of the measurement but introducing a source of artificiality into the research situation. Other systems increase the naturalness of the data collection activity by using equipment that allows the user to make head movements. With such equipment, head movements are discouraged because they are a source of error in eye tracking research; however, the benefits of ecological validity may outweigh the precision of more intrusive systems (Ellis et al., 1998).

ERICA

The Eye-gaze Response Interface Computer Aid (ERICA) system is one example of a relatively non-obtrusive eye-tracking technology. Currently the ERICA system is being used at the University of Virginia to examine the behavior of individuals with and without cognitive disabilities such as learning disabilities (LD) or Attention Deficit Hyperactivity Disorder (ADHD) using instructional materials delivered via the World Wide Web (www). We will provide a brief description of the way that the ERICA system collects information about eye movements to illustrate how eye-movement data can be collected.

The heart of the ERICA system consists of an infrared light-emitting diode (LED) centered in the lens of a camera placed below a computer monitor. A simplified diagram of the ERICA system appears in Fig. 1. The LED floods the user's face with a low level of infrared light that is reflected by two different

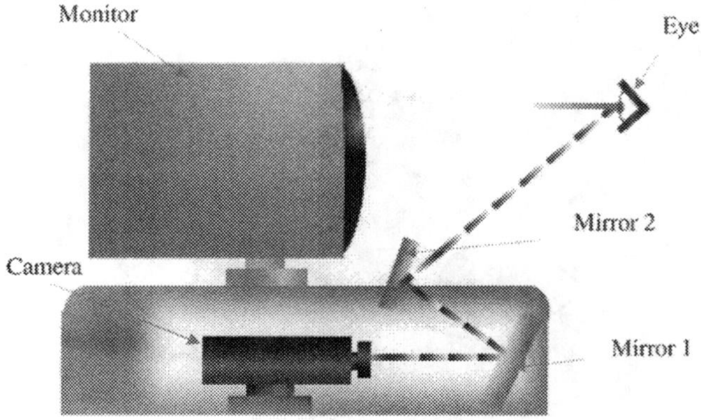

Fig. 1. The ERICA Hardware Setup.

features of the eye, the surface of the cornea and the retina. The corneal reflection is referred to as the glint. The retinal reflection is known as the bright eye. These reflections are collected by the camera that then feeds the information to a computer that calculates the position of the eye by comparing the relationship of the glint with the bright eye or center of the pupil. The ERICA system locates the glint and bright eye features of the camera image and determines where the user is looking based upon the separation between these two features, as shown in Fig. 2.

The ERICA system collects photographic information about the user's eye location at a rate of 60 times per second. This information is fed to a computer that algorithmically extracts the following information: (a) total time an individual spent viewing a given screen; (b) the time and location of all mouse movements and key presses; (c) the location of each eye fixation (places that the eye rests to collect information); (d) the duration of each eye fixation; (e) the direction and length of all eye movements; and (f) the diameter of the pupil at each fixation. Response latency can be collected by comparing the time that a visual or auditory prompt is presented by the computer and the time that the user executes an action with the mouse. A number of other eye tracking devices are available. Most generate similar data.

FUNDAMENTALS FOR UNDERSTANDING EYE-MOVEMENT RESEARCH

Currently, eye-movement researchers study visual behaviors to make inferences about cognitive and neurological processes (Morris & Rayner, 1991;

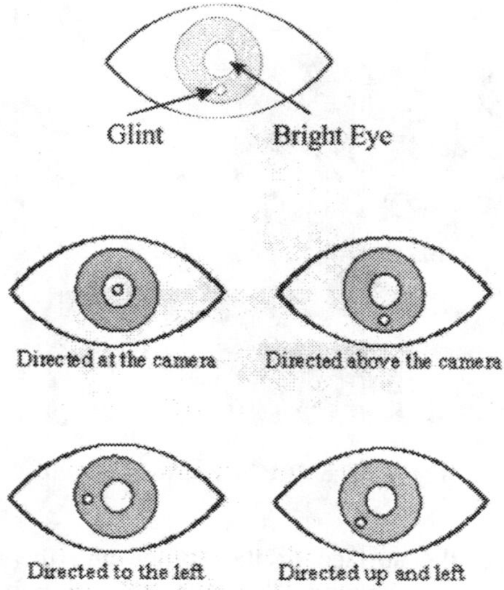

Fig. 2. Glint and Bright Eye Illustration.

Schiller, 1998). Although many aspects of the field remain highly controversial and different examinations of the same feature sometimes lead researchers to different conclusions (Kroll, 1992) several robust findings have emerged from eye-movement research. Several terms that are commonly used in eye-movement research are summarized in Table 1.

Major Findings of Eye Movement Research

Eye Movement Control is Developmental

Among the more robust findings in eye-movement research is the observation that the precision and control of eye movements develops across age (Ferrari, Kohler, Fogassi & Gallese, 2000; Richards & Hunter, 1998). As the neural mechanisms that control eye movements develop, infants are better able to exhibit selective attention. Thus, the ability to seek out visual information in the environment and make decisions about features of the environment that deserve attention appears to be, in part, learned or at least improved through practice.

Acuity Varies Across the Visual Field

The area of the retina associated with the most clear and useful vision is called the fovea. From the fovea, visual acuity degrades quickly along any axis toward

Table 1. Features and Definitions of Eye Movement Research.

Term	Definition
Anticipatory saccade	movements generated in anticipation of a target or stimulus in a particular location (Leigh & Zee, 1999).
Antisaccade	movements (based on instructions to do so) generated in the opposite direction to the sudden appearance of a target (Rafal, 1998).
Express saccade	very short latency saccades elicited when a novel stimulus is presented after a fixation stimulus has disappeared (Fischer, 1992).
Fixation	the time that the eyes rest on information to present it to the fovea. Most time spent in reading is devoted to fixations and only a small fraction is devoted to saccadic movements (Stark et al., 1991).
Fovea	the area of the retina with the highest resolution and greatest sensitivity to color (Groner, 1988).
Parafovea	the retinal area surrounding the fovea. Parafoveal density of receptors drops, reducing the acuity the visual field. Only general information such as word shape is extracted from the parafovea. Information about the physical features of text or pictures appearing in the parafoveal region, is believed to direct subsequent eye-movements (Lovegrove, 1991; Morris & Rayner, 1991).
Perceptual Span	the area of effective vision or the region from which useful information can be obtained (Morris & Rayner, 1991).
Pupilary Response	changes in pupil dilation associated with cognitive tasks but not with changes in light striking the eye (Backs & Walrath, 1992).
Reflexive saccade	movements generated to novel stimuli (visual, auditory, or tactile) that unexpectedly occur within the environment (Leigh & Zee, 1999).
Regression	small amplitude saccades that are directed backward relative to the conventional direction of reading. Regressions are believed to function to correct misreadings or to verify particular details. Generally, 5–20% of all saccades can be classified as regressions (Stark et al., 1991).
Return Sweep	large backward saccades that lead the eyes from the end of one line to the beginning of the next. (Stark et al., 1991).
Saccade	rapid eye movements performed to bring unidentified regions of text or picture into foveal vision for detailed analysis. The average saccade length is six to eight characters or about 2°. Saccades usually last 20–40 ms and no visual information is extracted from the stimulus while the eyes are moving (Morris & Rayner, 1991).
Smooth Pursuit	eye movement is stimulated by a visible moving target usually moving slower than $30° s^{-1}$. The ability of the eye to follow a moving target exists even early in infancy but at a lower level of performance. Smooth pursuit movements can be contrasted with reflexive and targeted saccades (Hainline, 1998).
Vergence	moves the eyes in opposite directions so that images of a single object fall simultaneously on both foveae (Leigh & Zee, 1999).

the periphery (Levy-Schoen, 1983). Although information about spaces and the presence of words in text and sources of light and shadows are available in extra-foveal vision, visual acuity is sharply limited to one to two degrees of the entire visual field (De Valois & De Valois, 1990). The size of a visual field is measured as the angle subtended at the eye by an object just filling it (Bruce & Green, 1990). For optimal vision of objects such as the words on this page, the images must be brought to the fovea and held there with a drift of less than 5°/second (Leigh & Zee, 1999).

The Visual World is Apprehended in Pieces

Only a small portion of the visual field yields clear visual information, therefore, it is necessary to collect images from the environment in small pieces. Movements of the eye, head and even the entire body are employed to carry out this task (Leigh & Zee, 1999). Irwin (1992) collected data suggesting that visual representations of the environment are built up from integration of several eye movements. Fisher (1992) noted that the information from these eye movements is not tied to absolute spatial position, suggesting that a more abstract code than the raw sensory data is employed to create long-lasting memories of the visual environment.

Attention in the Visual Field is Dynamically Allocated

Henderson (1992) suggested that attentional span was a better term to apply to discussions of visual information extracted from eye fixations. One reason for this suggestion is the observation that in text reading of English and other languages processed from left to right, useful information is extracted about three or four characters to the right of a fixation and about 14 characters to the left of a fixation (Pollatsek & Rayner, 1992). When reading Hebrew, however, this pattern of information extraction is reversed so that useful information is extracted about 14 characters to the right of a fixation and about three or four characters to the left. Visual attention is also accorded to different aspects of the visual field. For example, moving stimuli attract more attention than stationary flashing points in a field (Stein & Meredith, 1993). Additionally, Kinsler and Carpenter (1997) suggested that the pattern of saccades observed when individuals read music follows the flow of the music as it is being performed and not the visual pattern of information on the page.

Visual Attention Can Be Mediated by Cognitive Processes

Visual searches and reading behavior show different patterns of activity depending on the purpose and facility of the individual being observed (Haider & Frensch, 1999; Henderson, 1992; Hood, Atkinson & Braddick, 1998). If visual attention processes are "lower level" activities that are outside of volitional control, they should remain relatively constant across the life span. However, most tasks related to visual attention demonstrate an inverted U-shaped distribution with improvement in childhood and decline in old age (Enns, Brodeur & Trick, 1998; Ferrari et al., 2000). In general, the pattern of development in childhood reflects a shift from external control of attention (responding reflexively to visual stimuli) to more controlled and strategic allocation of attentional resources (Ruff, 1998). Both engaging attention on visual features and disengaging attention from highly salient stimuli are related to the development of effective visual attention in infants (Columbo & Janowsky, 1998).

Summary

Visual abilities develop across childhood and perhaps into adolescence. Because only a small part of the eye yields clear information, people must take in the visual environment in small units that are integrated to create a larger view of the environment. Attention within the visual span is dynamically allocated to perform different tasks (e.g. reading language displayed in different directions), therefore, visual attention can be mediated by higher-order cognitive processes. The flexibility of eye movements and visual attention activities and their responsiveness to cognitive mediation suggests that eye movements can indeed be used as clues to other less accessible processes such as cognition.

APPLICATIONS OF EYE-MOVEMENT TECHNOLOGY TO LEARNING PROBLEMS

Eye movements are often used to make inferences about underlying mental processes and their neural substrates (Leigh & Zee, 1999). For example, Gooding and Tallent (2001) compared eye movement and working memory data collected from individuals with either schizophrenia or bipolar disorder and found different patterns of performance for the two groups. Their results suggested that the two disorders affect different cognitive and neural functions. A number of other eye-movement studies are to be found in the psychiatric literature. Most of these studies assume that eye movement differences are the result and not the cause of differences in underlying processes. When considering individuals with disorders that are more common in school-aged populations (e.g. LD & ADHD), researchers have not always made such a distinction.

Although the relationship between eye-movement abnormalities and reading difficulties is unclear, individuals with certain occulomotor abnormalities do indeed have difficulty reading (Leigh & Zee, 1999). Researchers in fields related to special education (e.g. Pavlidis, 1990, 1991) have often assumed that eye-movement deficits were the cause rather than the result of certain problems. For most individuals with reading problems, the data do not support such assumptions. Rather, it appears that eye-movement differences in cognitive tasks such as reading are the result of the individual's facility with the skill of reading (Rayner, 1985a, b). Most readers begin to exhibit patterns of behavior that are similar to individuals with LD or other mild disabilities when the demands of the cognitive task presented to them become as proportionately difficult for them as they are for individuals with disabilities.

Reading problems are perhaps the most often examined aspect of learning difficulties that is studied with eye-movement technology. For most individuals, there is clearly a sensible link between eye-movement behavior and reading because the first stage of reading is the acquisition of information through the eyes. Reading is a complex, multi-step process that should be viewed as containing at least two stages (McConkie, Reddix & Zola, 1992). In the first stage, the reader acquires information from the printed page. In the second stage, the reader translates the visual information to language and carries out higher order, information-processing activities. Interference with the visual decoding of text (e.g. briefly blurring a word on a computer screen) results in a different pattern of behavior than interference with the language processes of reading (e.g. inserting a phonetically regular pseudoword in a text). Specifically, disruption of visual acquisition results in a consistent pattern of delay of saccades while disruption of the language processes results in a highly inconsistent pattern of saccadic delays (McConkie et al., 1992). In addition to supporting the distinction between visual and linguistic information-processing in reading, these findings suggest that eye movements are related to the individual's general language abilities and in many tasks, such as reading about specific content areas, also related to one's store of prior knowledge. The study of eye movements in reading is further complicated by the observation that skilled readers do not jump neatly from word to word but sometimes skip short words altogether and fixate on long or more important words more than once (Sereno, 1992). The extent to which readers actually skip words in text is controversial. It may be that skilled readers make much longer saccades around highly familiar words such as "the" rather than skipping the word entirely (O'Regan, 1992).

The foregoing discussion highlights some of the difficulties in conducting eye-movement research relative to the actual process of reading. In addition to the complexity of the reading process itself, eye-movement research is also complicated by difficulties in sample selection for populations exhibiting difficulties in school-learning tasks. Kavale and Nye (1985–1986) demonstrated that there is a complex interaction of achievement, linguistic, neuropsychological, and social/behavioral domains within the population of school-identified students with LD. No one domain was predominant in the group. The problem is further complicated by the presence of false positives in the group considered to have a disability and the presence of false negatives within the group of typical children selected for any study (Brigham, Tochterman & Brigham, in press). Therefore, positive findings of differences among groups of typical children and children with disabilities may not generalize well across children with similar diagnostic labels. Conversely, failure to detect differences among groups may be the result of sampling error rather than truly non-significant differences among the groups. With these cautions in mind, we shall now describe some of the recent research on eye movements conducted with children and youths with disabilities.

Saccadic Eye Movements and Dyslexia

Dogen and Pavlidis (1990) compared responses of 39 male students with dyslexia and at least a two-year delay in reading to 35 typical student controls. A series of dots appeared and were then extinguished on a computer screen from left to right. Participants were required to fixate upon each dot and execute a bar press on the keyboard. All dots were 4° apart and appeared for one second each. After the entire series of dots had been displayed, an audible signal cued the subjects to look at each point where a dot would have appeared on the blank screen and to do so at the time that it would have appeared. Interestingly, the spatial accuracy of the groups was not significantly different; however, the timing with which the dyslexic group executed their eye movements was significantly worse and less regular (rhythmic) than the control participants.

Fischer and Weber (1990) collected data on the saccadic movements of 20 children and youths with dyslexia, 17 typical children and youths, and 15 typical adults. All participants focused on a central fixation point displayed on a computer screen and moved their eyes to a target point appearing elsewhere on the screen. Dyslexic children produced more express saccades than did their age-matched controls. The authors suggested that the results obtained were more likely the product of defects in the system of visual attention and its control over the oculomotor system rather than defects in the oculomotor system itself.

Shapiro, Ogden and Lind-Bland (1990) presented short and long words in isolation on a computer screen to participants with and without dyslexia whose eye movements were being monitored. The dependent measure was the ability of the participants to identify the stimulus word. Dyslexic students were significantly less able to identify long words but equally accurate in identification of short words with the control students. While long words required more than one fixation for decoding the word, there were no differences reported in the number of eye movements made by each group.

Reddington and Cameron (1991) examined several aspects of eye-movement behavior in dyslexic and non-dyslexic children aged between 7.25 and 10.25 years. Only saccadic length measures attained significance in between group analyses. In a sentence-reading task, dyslexic children were found to make shorter saccadic movements than did children with adequate reading abilities.

Biscaldi, Fischer and Stuhr (1996) compared individuals who made large numbers (>30%) of express saccades (ES) with individuals who had typical levels of ES. The ES makers included six dyslexic males, two non-dyslexic males and two non-dyslexic females. The non-ES maker group contained one dyslexic male, six non-dyslexic males, and three non-dyslexic females. Participant ages ranged from 12 to 32 years. All participants completed five different experimental sessions. In standard saccade conditions, participants were presented with a central fixation stimulus and told to look quickly at a subsequently appearing target. Antisaccade conditions required participants to look to the side of a fixation stimulus opposite to the subsequently presented target. A memory-guided saccade task presented a briefly flashed target stimulus while the participant was focused on a different fixation point. After the stimuli were removed, the participant was required to look at the remembered location of the target. The ES makers exhibited significantly more express saccades (>30%) than the non-ES makers (<15%). In addition to the high numbers of ES's, the ES makers demonstrated significantly greater difficulty in suppressing reactive glances in the direction of the target during the antisaccade condition as well as in maintaining focus on the fixation point during the memory task. Biscaldi, Fischer and Stuhr suggested that the differences between groups were probably related to the tendency of ES makers' to react reflexively instead of purposively to visual stimuli.

Biscaldi, Gezeck and Stuhr (1998) compared the saccadic movements of 92 children with dyslexia according to the *International Classification of Diseases* (ICD-10) (World Health Organization, 1992). Saccadic eye movements were measured in two experimental sessions that were each preceded by a number of practice sessions. Two tasks were included in the study, a single target task and a sequential targets task. The single target task was always presented first.

In the single target task, a fixation point was presented on a computer monitor 1.7 seconds before a target stimulus was presented to either the left or right of the fixation stimulus. Both stimuli were extinguished after the target stimulus had appeared for 800 ms. Participants were to fixate on the target stimulus as quickly as possible. In the sequential targets task, a fixation point appeared on the left side of the computer monitor for 1.2 seconds. A series of four targets appeared on a horizontal line, left to right, one per second in sequence. The subjects were to fixate on each target in sequence as quickly as possible. Dyslexic readers were distinguished from non-dyslexic controls by increased mean and standard deviations of saccadic reaction times (latency in response to a target stimulus) and the count of late saccades. Additionally, dyslexic subjects made significantly more regressions than did control subjects. Fifty percent of the dyslexic group was found to exhibit atypical saccadic movements as compared to 20% of the control group. The number of express saccades approached but did not attain significance, with the dyslexic group making the larger number of these movements. Finally, Biscaldi, Gezeck and Stuhr suggested that the development of saccadic control for the dyslexic group appeared to be on a similar trajectory to the control group, but progressing at a much slower rate.

Summary

Compared to typical readers, children with dyslexia tend to make more express saccades and have problems with both the timing and accuracy of their eye movements. Most of these results have been collected in relation to non-linguistic stimuli to avoid the effects of reading ability on eye movements. Some authors have suggested that the eye-movement differences observed in people with dyslexia may be the result of attentional processes rather than dysfunction in the oculomotor system itself. The exact relation of these phenomena to reading ability, however, remains unclear.

Vergence Movements and Reading Ability

Fowler, Riddell and Stein (1990) provided eye movement data from 80 typical students aged 4–6 to 6–6 and 82 dyslexic students ages 6–0 to 18–0 regarding their ability to control vergence movements and localize small objects in space. Children with low scores on a test of vergence ability had significantly more difficulty in localizing objects in space in either visual field. but tended to make more errors in the left visual field (right hemisphere controlled) than in the right visual field (left hemisphere controlled). The authors interpreted their results to suggest that many children with reading disabilities may be characterized by deficits in behaviors and processes normally controlled by the right cerebral hemisphere.

Riddell, Fowler and Stein (1990) reported a similar experiment in which 131 children were classified as having poor or adequate vergence control and poor or adequate reading abilities. The participants were presented with a central fixation dot on a computer screen that was extinguished just before a target dot was presented for 200 ms elsewhere on the screen. Subjects were requested to point in the direction that the target dot had appeared to move relative to the initial, fixation dot. A clear developmental trend was found for both vergence control and pointing accuracy in both groups. Participants with lower vergence control ratings made significantly more pointing errors than did participants with adequate vergence control ratings. Significantly more errors were made in the left visual field than the right visual field, again suggesting children with poor vergence control may be characterized by dysfunctions in right cerebral hemispheric functions.

Summary

Vergence movements bring objects onto both foveae simultaneously. They are thought to be important for locating objects in space. Some evidence exists to indicate that for a small group of children, vergence movements present a particular challenge. Further, these children exhibit significant difficulties in locating objects in space. While the studies presented here do not establish a clear link between problems in vergence movement ability and difficulties in finding and keeping one's place on a page of written text, they present at least a plausible explanation for the difficulties experienced by some students.

Saccadic Eye Movements and Other Reading Difficulties

Fletcher (1991) collected eye movement data from 10 students with reading disabilities and 10 typical students while they read sentences in isolation on a computer screen. The stimulus sentences were categorized as: (a) syntactically ambiguous, (b) semantically anomalous, and (c) regular or control sentences. The syntactic and semantic irregularities were included to induce comprehension errors in the readers. The results suggested that both groups of students employed error recovery (ER) patterns that were organized, strategic, and efficient according to an a priori classification scheme developed by the author. ER patterns were similar between the two groups with the exception of restarts and disorganized movement patterns that were both exhibited more frequently by disabled readers.

Fletcher (1990) presented individual sentences with semantic or syntactic errors (high demand condition) and non-errant sentences (low demand condition) to 25 students with comprehension disabilities and 35 typical students and recorded their eye movements as they read. Both groups of students exhibited highly variable patterns of eye movements with the control groups demonstrating the most variability in the low demand condition. Fletcher conjectured that the presence of errant sentences may have made some readers suspicious and prompted them to examine all sentences more closely. It is interesting that the control group exhibited this differential response to the nonerrant condition. Failure to demonstrate this increased variability may reflect the absence of active engagement by the disabled group.

Fletcher (1993) obtained eye-movement data from 17 adolescents with reading disabilities (comprehension scores more than 2 years behind grade level) and 15 control subjects completing a sentence-reading task. The saccade direction, saccade length and fixation durations were then translated into musical code. Disabled readers were found to be less rhythmic and have more variability in the pitch in the musical code translated from their eye movements.

Eden, Stein, Wood and Wood (1995) examined measures of phonological awareness, verbal memory, and object naming along with eye-movement data fixation control, vertical tracking ability, horizontal tracking, vergence ability, and depth perception that were obtained from students who were classified as reading disabled (RD), reading backwardness (RB), and a group of typical controls, all from grades 3 and 5. The children with RD and RB demonstrated significant difficulties in both verbal and visual tasks significantly more often than did typical control students. Sixty-eight percent of the variance in reading ability could be predicted by combining the visual and verbal scores in a multiple regression analysis. Eden, Stein, Wood, and Wood concluded that the earlier decision to discount the contribution of dysfunctional visual and oculomotor systems to reading problems was premature.

Summary

Studies of students with reading disabilities that are more generally defined than dyslexia demonstrate a less clear pattern of eye-movement behavior than do the children in the studies where more restrictive definitions were employed. Also, the stimuli employed with the generally defined reading disability groups included more linguistic tasks than those employed with children identified as dyslexic. When various linguistic measures (e.g. phonetic awareness) were considered in conjunction with eye movement-data, the explanatory power greatly exceeded that of either component.

Integration

Rayner's (1985b) paper clearly demonstrated that the majority of reading problems could be explained without resorting to eye-movement factors. Despite the evidence that continues to accrue that some groups of children exhibit eye movements that are quite dissimilar to their more acceptably performing peers, there is little reason to doubt the conclusions about eye movements offered by Rayner more than 15 years ago. Rather than a causal factor involved in unsatisfactory performance of reading or other skills involving visual perception, eye movements may be better understood as a manifestation of other, underlying cognitive processes. Such a conclusion has important implications for the role of eye-movement data in treatment of students with various performance deficits. Earlier considerations for eye-movement differences suggested that direct training of eye movements would enable the individual to execute other cognitive processes such as reading. A more realistic view may be that because eye movement differences are the result of cognitive processes such as selective attention, training in the underlying process may yield changes in the eye-movement behavior of the individual. For example, evidence of the success of a reading comprehension strategy might involve demonstration that students with poor comprehension pretest scores actually spent more time looking at critical information in texts after training. Another application that grows out of this reconsideration of the eye-movement data is the extent to which text-embedded strategies can be shown to actually prompt learners to engage in behaviors related to learning outcomes (e.g. examination of graphics included to illustrate and disambiguate key concepts). We will conclude this paper with a brief discussion of the Visual Education for Scientific Literacy (VESL) project presently being conducted along this line of reasoning at the University of Virginia.

VESL

For the past few months we have been collecting data regarding learner behavior on web-based instructional tools using the previously described ERICA system. We proposed this project in response to the expanding endorsement of academic uses of the www by educators and politicians. Our initial questions were the extent to which students with and without cognitive disabilities would exhibit different patterns of eye movements while working with a web site regarding information from the Virginia Standards of Learning. To date, we have detected few reliable differences in eye-movement behavior between these groups of students. Three consistent findings have emerged, however. First, most students exhibit reading-like behavior even if they are unlikely to actually be reading the website. An extreme case example may help to clarify the last statement. Most people decode words in a matter of milliseconds. For the least able readers who have helped us in this project, we have observed fixation times of up to 5 seconds per word. In fact, in the most extreme case we observed, the student actually used the mouse to point at each word in the text. He then fixated on the word from one to five seconds. His pattern of movements was actually more predictable than the better readers who worked with the same website. It is very unlikely that an individual reading in such a slow and laborious manner is actually extracting meaning from his efforts despite the regularity of the eye movements.

Our second consistent finding to date is that none of our participating students actually look at non-text features of the website such as tables of data, hyperlink bars, illustrations or animations. We are not particularly surprised at the limited attention paid to these features because the text in the display we are currently using never explicitly refers to any of the graphics. However, we were surprised that even the individual whose reading behavior was previously described spent proportionately the same amount of time looking at each block of text and each graphic as the other, more adept readers we have observed.

Finally, the results of a post-test of the website content that we created suggest that none of the participants actually retained much of the information from the website. Given the instructions to read the website in any manner that they wished to prepare for the post-test that would immediately follow the reading, most students simply made a single reading pass through each of the text section, glanced at an animation demonstrating the phenomenon under consideration and ignored a table of data showing the trends of the phenomenon over time. Average performance on the post-test was a disappointing 30 to 40%. These results suggest that simply providing students with websites and hoping that they will explore them and retain important and substantial amounts of information is a naïve approach to instruction, at least with the present state of website development.

Rather than simply bemoaning the apparent absence of efficacy of web-delivered instruction for some learners, we are working to modify ineffective websites that present important content to make them more memorable and profitable for school-based learning tasks. We believe that improvement in traditional measures of instructional effectiveness such as achievement tests will be important validations of our efforts. We are also examining the extent to which our website modifications result in different eye-movement behaviors on the part of our participating students. If changes in outcome measures can be clearly related to altered patterns of visual inspection of the website, then it may truly be concluded that "the eyes have it."

ACKNOWLEDGMENTS

We express our sincere appreciation to Thomas Hutchinson and Chris Lankford of ERICA, Inc. for their unflagging support of this project. We also wish to acknowledge the support of Lisa McConnell and the staff and students of Elk Hill Farms as well as Mr. Earl Pappy and the staff and students of Boushall Middle School. This work was partially supported by U.S. Department of Education Grant 5–34333.

REFERENCES

Backs, R. W., & Walrath, L. C. (1992). Eye movement and pupilary response indices of mental workload during visual search of symbolic displays. *Applied Ergonomics*, *23*, 243–254.

Biscaldi, M., Fischer, B., & Stuhr, V. (1996). Human express saccade maker are impaired at suppressing visually evoked saccades. *Journal of Neuropsychology*, *76*(1), 199–214.

Biscaldi, M., Gezek, S., & Stuhr, V. (1998). Poor saccadic eye movement correlates with dyslexia. *Neuropsychologia*, *36*(11), 1189–1202.

Brigham, F. J., Tochterman, S., & Brigham, M. S. P. (in press). Students with emotional and behavioral disorders and their teachers in test-linked systems of accountability. *Diagnostique*.

Bruce, V., & Green, P. R. (1990). *Visual perception: Physiology, psychology, and ecology* (2nd ed.). Hillsdale, NJ: Lawrence Erlbaum.

Columbo, J., & Janowsky, J. S. (1998). A cognitive neuroscience approach to individual differences in infant cognition. In: J. E. Richards (Ed.), *Cognitive Neuroscience of Attention* (pp. 363–392). Mahwah, NJ: Lawrence Erlbaum Associates.

De Valois, R. L., & De Valois, K. K. (1990). *Spatial vision*. New York: Oxford University Press.

Dogen, C. E., & Pavlidis, G. T. (1990). Sequential, timing, rhythmic and eye movement problems in dyslexics. In: G. T. Pavlidis (Ed.), *Perspectives of Dyslexia* (Vol. 1, pp. 221–251). New York: John Wiley and Sons.

Eden, G. F., Stein, J. F., Wood, M. H., & Wood, F. B. (1995). Verbal and visual problems in reading disability. *Journal of Learning Disabilities*, *28*(5), 272–290.

Ellis, S., Candrea, R., Misner, J., Craig, C. S., Lankford, C. P., & Hutchinson, T. E. (1998). Windows to the soul? What eye movements tell Us about software usability. Unpublished manuscript. The University of Virginia: Charlottesville, VA.

Enns, J. T., Brodeur, D. A., & Trick, L. M. (1998). Selective attention of the life span: Behavioral measures. In: J. E. Richards (Ed.), *Cognitive Neuroscience of Attention* (pp. 393–418). Mahwah, NJ: Lawrence Erlbaum Associates.

Ferrari, P. F., Kohler, E., Fogassi, L., & Gallese, V. (2000). The ability to follow eye gaze and its emergence during development in macaque monkeys. *Proceedings of the National Academy of Science: United States of America*, *97*(25), 13997–14002.

Fischer, B. (1992). Saccadic reaction time: Implications for reading, dyslexia, and visual cognition. In: K. Rayner (Ed.), *Eye Movements and Visual Cognition: Scene Perception and Reading* (pp. 31–45). New York: Springer-Verlag.

Fischer, B., & Weber, H. (1990). Saccadic reaction times of dyslexic and age-matched normal subjects. *Perception, 19*, 805–818.

Fisher, D. L. (1992). The matter and mechanisms of integration. In: K. Rayner (Ed.), *Planning and Producing Saccadic Eye Movements* (pp. 192–202). New York: Springer-Verlag.

Fletcehr, J. (1991). Qualitative descriptions of error recovery patterns across reading level and sentence type: An eye movement analysis. *Journal of Learning Disabilities, 24*(9), 568–575.

Fletcher, J. (1993). Eye-movement rythmicity and reading comprehension. *Journal of Learning Disabilities, 26*, 683–688.

Fletcher, J. M. (1990). Component skill comparisons across reading level and processing demand. *Reading Psychology, 11*, 193–239.

Fowler, M. S., Riddell, P. M., & Stein, J. F. (1990). Vergence eye movement control and spatial discrimination in normal and dyslexic children. In: G. T. Pavlidis (Ed.), *Perspectives of Dyslexia* (Vol. 1, pp. 253–273). New York: John Wiley and Sons.

Gooding, D. C., & Tallent, K. A. (2001). The association between antisaccade task and working memory task performance in schizophrenia and bipolar disorder. *Journal of Nervous & Mental Disease, 189*(1), 8–16.

Groner, R. (1988). Eye movements, attention and visual information processing: Some experimental results and methodological considerations. In: G. Luer & U. Lass & J. Shallo-Hoffmann (Eds), *Eye Movement Research: Physiological and Psychological Aspects* (pp. 295–319). Toronto: C. J. Hogrefe.

Haider, H., & Frensch, P. A. (1999). Eye movement during skill acquisition: More evidence for the information-reduction hypothesis. *Journal of Experimental Psychology, 25*(1), 172–190.

Hainline, L. (1998). Eye movement, attention and development. In: J. E. Richards (Ed.), *Cognitive Neuroscience of Attention* (pp. 163–178). Mahwah, NJ: Lawrence Erlbaum Associates.

Heller, D. (1988). On the history of eye movement recording. In: G. Luer & U. Lass & J. Shallo-Hoffman (Eds), *Eye Movement Research: Physiological and Psychological Aspects* (pp. 37–51). Toronto, Canada: C. J. Hogrefe.

Henderson, J. M. (1992). Visual attention and eye movement control during reading and picture viewing. In: K. Rayner (Ed.), *Planning and Producing Saccadic Eye Movements* (pp. 260–283). New York: Springer-Verlag.

Hood, B. M., Atkinson, J., & Braddick, O. J. (1998). Selection-for-action and the development of orienting and visual attention. In: J. E. Richards (Ed.), *Cognitive Neuroscience of Attention* (pp. 219–286). Mahwah, NJ: Lawrence Erlbaum Associates.

Irwin, D. E. (1992). Visual memory within and across fixations. In: K. Rayner (Ed.), *Planning and Producing Saccadic Eye Movements* (pp. 146–165). New York: Springer-Verlag.

Kavale, K. A., & Nye, C. (1985–1986). Parameters of learning disabilities in achievement, linguistic, neuropsychological, and social/behavioral domains. *Journal of Special Education, 19*, 443–458.

Kinsler, V., & Carpenter, R. H. S. (1997). Saccadic eye movements while reading music. *Vision Research, 35*, 1447–1458.

Kroll, J. F. (1992). Making a scene: The debate about context effects for scenes and sentences. In: K. Rayner (Ed.), *Planning and Producing Saccadic Eye Movements* (pp. 284–292). New York: Springer-Verlag.

Leigh, R. J., & Zee, D. S. (1999). *The neurology of eye movements* (3rd ed.). New York: Oxford University Press.

Levy-Schoen, A. (1983). Central and peripheral processing. In: R. Groner, C. Menz, D. F. Fisher & R. A. Monty (Eds), *Eye Movements and Psychological Functions: International Views* (pp. 65–71). Hillsdale, NJ: Lawrence Erlbaum Associates.

Lovegrove, W. (1991). Spatial frequency processing in dyslexic and normal readers. In: J. F. Stein (Ed.), *Vision and Visual Dysfunction: Vision and Visual Dyslexia* (Vol. 13, pp. 148–154). Boca Raton, FL: CRC Press.

McConkie, G. W., Reddix, M. D., & Zola, D. (1992). Perception and cognition in reading: Where is the meeting point? In: K. Rayner (Ed.), *Planning and Producing Saccadic Eye Movements* (pp. 293–303). New York: Springer-Verlag.

Morris, R. K., & Rayner, K. (1991). Eye movements in skilled reading: Implications for developmental dyslexia. In: J. F. Stein (Ed.), *Vision and Visual Dysfunction: Vision and Visual Dyslexia* (Vol. 13, pp. 233–242). Boca Raton, FL: CRC Press.

O'Regan, J. K. (1992). Optimal viewing position in words and the strategy-tactics theory of eye movements in reading. In: K. Rayner (Ed.), *Planning and Producing Saccadic Eye Movements* (pp. 333–354). New York: Springer-Verlag.

Pavlidis, G. T. (1990). Detecting dyslexia through opthalmo-kinesis: A promise for early diagnosis. In: G. T. Pavlidis (Ed.), *Perspectives of Dyslexia* (Vol. 1, pp. 199–220). New York: John Wiley and Sons.

Pavlidis, G. T. (1991). Diagnostic significance and relationship between dyslexia and erratic eye movements. In: J. F. Stein (Ed.), *Vision and Visual Dysfunction: Vision and Visual Dyslexia* (Vol. 13, pp. 263–270). Boca Raton, FL: CRC Press.

Pollatsek, A., & Rayner, K. (1992). What is integrated across fixations? In: K. Rayner (Ed.), *Planning and Producing Saccadic Eye Movements* (pp. 166–191). New York: Springer-Verlag.

Rafal, R. (1998). The neurology of visual orienting: A pathological disintegration of development. In: J. E. Richards (Ed.), *Cognitive Neuroscience of Attention* (pp. 181–218). Mahwah, NJ: Lawrence Erlbaum Associates.

Rayner, K. (1985a). Do faulty eye movements cause dyslexia? *Developmental Neuropsychology*, *1*(1), 3–15.

Rayner, K. (1985b). The role of eye movements in learning to read and reading disability. *Remedial and Special Education*, *6*(6), 53–60.

Rayner, K. (1992). Eye movements and visual cognition: Introduction. In: K. Rayner (Ed.), *Eye Movements and Visual Cognition: Scene Perception and Reading* (pp. 1–7). New York: Springer-Verlag.

Reddington, J. M., & Cameron, K. D. (1991). Visual and auditory information processing in dyslexia: The possibility of subtypes. *International Journal of Disability, Development, and Education*, *38*(2), 171–203.

Richards, J. E., & Hunter, S. K. (1998). Attention and eye movement in young infants: Neural control and development. In: J. E. Richards (Ed.), *Cognitive Neuroscience of Attention* (pp. 131–162). Mahwah, NJ: Lawrence Erlbaum Associates.

Riddell, P. M., Fowler, M. S., & Stein, J. F. (1990). Spatial discrimination in children with poor vergence control. *Perceptual and Motor Skills*, *70*, 707–718.

Ruff, H. A. (1998). Selective attention: Its measurement in a developmental framework. In: J. E. Richards (Ed.), *Cognitive Neuroscience of Attention* (pp. 419–425). Mahwah, NJ: Lawrence Erlbaum Associates.

Schiller, P. H. (1998). The neural control of visually guided eye movements. In: J. E. Richards (Ed.), *Cognitive Neuroscience of Attention* (pp. 3–50). Mahwah, NJ: Lawrence Erlbaum Associates.

Sereno, S. B. (1992). Early lexical effects when fixating a word in reading. In: K. Rayner (Ed.), *Planning and Producing Saccadic Eye Movements* (pp. 304–316). New York: Springer-Verlag.

Shapiro, K. L., Ogden, N., & Lind-Bland, F. (1990). Temporal processing in dyslexia. *Journal of Learning Disabilities, 23*(2), 99–107.

Stafford, B. M. (1994). *Artful science: Enlightenment entertainment and the eclipse of visual education*. Cambridge, MA: MIT Press.

Stark, L. W., Giveen, S. C., & Terdiman, J. F. (1991). Specific dyslexia and eye movements. In: J. F. Stein (Ed.), *Vision and Visual Dysfunction: Vision and Visual Dyslexia* (Vol. 13, pp. 203–232). Boca Raton, FL: CRC Press.

Stein, B. E., & Meredith, M. A. (1993). *The merging of the senses*. Cambridge, MA: MIT Press.

World Health Organization. (1992). *The ICD-10 classification of mental and behavioural disorders: Clinical descriptions and diagnosis guidelines*. Geneva, Switzerland: Author.

THE ROLE OF ASSISTIVE TECHNOLOGIES AND EMERGING TECHNOLOGIES IN DEVELOPING LITERACY SKILLS FOR STUDENTS WITH DISABILITIES

Tara Jeffs

ABSTRACT

What exactly does the research tell us about the literacy process for students with disabilities, the implementation of computers, and the role of assistive technology and emerging technologies (i.e. Internet, and Electronic Performance Support Systems) in developing literacy skills in these students? This chapter discusses: (1) the use of instructional technology and assistive technology that have been successfully used over the past decade to build literacy skills for students with disabilities, (2) contemporary educational research of the Internet as it relates to issues of effectiveness, design, and individual differences, and (3) highlights from a recent research study involving parents and children using assistive technology in a literacy experience on the Internet.

Technological Applications, Volume 15, pages 61–94.

ISBN: 0-7623-0815-x

INTRODUCTION

The ability to read and write allows an individual to gain independence both in and out of the classroom. Reading is critical to success in our changing communities. Children who become adults with poor levels of literacy are at an increasing disadvantage in a society that is creating even higher demands for effective reading and writing skills in the workplace (Torgesen, 2000).

The National Institute of Child Health and Human Development (NICHD, 1999) have revealed recent research findings concerning the area of reading development, reading disorders and the importance of literacy skills. Research findings reveal that the illiteracy rate in the United States is unacceptably high (NICHD, 1999). "Converging evidence from longitudinal, population-based data indicate that at least 17% to 20% of school-age children have a significant reading disability" (NICHD, 1999, p. 2).

CHARACTERISTICS OF STUDENTS WITH READING DIFFICULTIES

Reading difficulty is one of the most significant problems experienced by children identified with learning disabilities (Swanson, 1999). By identifying general characteristics of students with reading difficulties, interventions can be implemented to assist or overcome such reading barriers. Prevalence studies and assessments conducted by the NICHD (1999) found:

(1) While public schools identify approximately four times as many boys as girls as reading disabled, NICHD longitudinal & epidemiological studies show that as many girls as boys have difficulties learning to read;
(2) Difficulties learning to read do not reflect a transient developmental lag but rather NICHD longitudinal studies indicate that of children who are reading disabled in the third grade, 74% remain disabled at the end of high school (p. 5).

Reading research currently addresses three specific aspects of the reading process: word recognition, reading comprehension, and motivation. These areas of research will be discussed briefly, in order to collectively gather a broader understanding of the difficulties individuals with disabilities encounter in the literacy process.

"The most frequent characteristics observed among children and adults with reading disabilities is a slow, labored approach to "decoding" or "sounding-out" unknown or unfamiliar words" (NICHD, 1999, p. 5). "Oral reading is also hesitant and characterized by frequent starts and stops and multiple mispronunciations" (NICHD, 1999, p. 5). As a result, common reading difficulties include problematic recognition of words by sight, and inability to productively use phonetic cues to help decode the unfamiliar word (Torgesen, 2000). The NICHD research reveals that phonemic awareness skills assessed in kindergarten, in addition to assessment of the child's ability to provide letter and number names and letter sounds, are strong predictors of difficulties learning to read.

A powerful predictor of reading comprehension is the speed and accuracy of reading single words that in return relates back to phonemic awareness skill development. Mastropieri and Scruggs (1997) substantiate students with learning disabilities typically exhibit deficits in reading comprehension, which may include summarizing the main idea, facts, and details of text material, but also in the interpreting and making inferences about the information presented.

Motivation is another factor in reading development and reading disorders. The NICHD (1999) findings suggest, "the amount of improvement a reader with disabilities may gain in learning to read is highly related to their willingness to persist despite difficulties" (p. 6). Baker and Wigfield (1999) state that because reading is an activity that children can choose to do or not to do, it requires motivation. Baker and Wigfield conducted a study involving 140 fifth graders and 230 sixth graders validating that motivation is multifaceted. Research evidence shows correlations between the motivation scales and the achievement measure differ with age, gender, and ethnicity.

CHARACTERISTICS OF STUDENTS WITH WRITING DIFFICULTIES

Students with learning disabilities experience difficulty in all stages of the writing process (Lewis, Graves, Ashton & Kieley, 1998). In comparison to normally achieving peers, students with disabilities tend to write a shorter, less coherent, and less refined writing product as a result of the inability to generate ideas, organize text, and apply metacognitive skills (Englert & Raphel, 1988; Graham, Schwartz & MacArthur, 1993; McAlister, Nelson & Bahr, 1999; Newcomer, Barenbaum & Nodine, 1988). Students with disabilities make considerably more spelling, capitalization, and punctuation errors; commonly referred to as writing mechanics (Graham, Schwartz & MacArthur, 1993). MacArthur (1999) states "problems with writing mechanics interfere with higher level composing processes and affect both the quantity and overall quality of writing" (p. 170). For some students with disabilities, the physical demand of handwriting produces a slow and less legible writing product (Graham & Weintraub, 1996).

The importance of examining basic characteristics of students with disabilities in the literacy process enables the production of new delivery and teaching models. Such models include the use of technology. These models are continuously being examined to determine their effectiveness in assisting students in overcoming the barriers involved in the literacy process.

THE ROLE OF TECHNOLOGY IN LITERACY DEVELOPMENT

Meyer and Rose (1998) differentiate the use of computers from other technologies that can contribute to the teaching of reading by stating, "More than any other intervention, computers are multipurpose devices that perform different functions, essentially becoming different machines through the use of varied software" (p. 8).

Versatility and the ability to customize computers provide flexibility to do many things in many ways, and have the capability to create appropriate learning environments for every student. Computers offer students learning environments that encourage confidence, enthusiasm, and enjoyment, all of which support learning (Meyer & Rose, 1998).

As technology becomes more prevalent in our daily lives both on and off the job and begins to transform our roles and responsibilities in today's society, the general public is exerting pressure for comparable changes within our schools (http://www.ed.gov/pubs/EdReformStudies/TechReforms). Many states, as part of their Goals 2000 planning effort, are directed to develop technology plans defining how they will implement technology to support systemic reform and to help their students achieve high standards (http://www.ed.gov/pubs/EdReformStudies/TechReforms).

In today's classrooms, students are using computers for drill and practice, simulated problem-solving, word processing, Internet searches, e-mail correspondence with subject matter experts, professionals at universities and government agencies (i.e. NASA, USDA), and collaborating with other schools across the nation and around the world (i.e. Cyber Olympics, Stars, and CoVis).

Instructional Technology

In order to optimize technological capabilities, we need to take a closer look at "what technologies do best: process, present, store, and retrieve information and images on demand" (Hannafin, Hannafin, Hooper & Rieber, 1996, p. 396). In comparison to what humans do best: problem-solve, concept formation, and mental processing (Ertmer & Newby, 1993). Educational theorists, such as Rousseau, Dewey and Bruner, viewed learning in the activity of the learner. Technology can provide hands-on learning, situated learning, inquiry-based learning, and discovery learning that make up the framework for inquiry and interpersonal interaction needed for learning.

"One impact that technology may have on learning is that it can lighten the cognitive load of the learner by allowing the learner to attend to higher-level thinking skills by "off-loading" basic cognitive demands" (Duffy & Cunningham, 1996, p. 187). For example, a database may ease the need for memorization of facts or a word processor may permit a writer to express ideas freely without getting bogged down by the basic mechanics of spelling.

Lewis, Graves, Ashton and Kieley (1998) conducted research involving 132 students (average grade placement 6.4) with learning disabilities. Their findings revealed that word prediction programs enhance the text entry speed of most students by allowing the student to simply recognize a word instead of spelling it out by individual letters. Technology can provide a powerful and responsive environment that involves the learner in the learning process and gives consideration for individual learning styles, motivations, and goals (Cooper, 1993).

Hypermedia

Many educators and literacy scholars are re-examining the learning experience and the education process as a result of advancing hypermedia technologies (Dillion & Gabbard, 1998). According to Jonassen et al. (1997):

> Hypertext and hypermedia technology provides the learner with two critical features of learning environments identified by Hannafin (1992), the process of linking new knowledge to old and modifying and enriching existing knowledge and increasing the number of access points to information (p. 120).

Although they vary greatly, both of these technologies engage the learner in multiple representations of content or information to be learned. Within the hypertext or hypermedia learning environments the shift of control moves away from the typical software program to the user, so that he or she makes the decision about what happens next (Lewis, 1998). Thus providing the learner an opportunity to engage and take charge of his/her own learning experience.

Higgins, Boone and Lovitt (1996) investigated the use of hypermedia study guides and information retention with students identified as learning disabled and students receiving remedial services. The study sample included 13 students with learning disabilities and 12 remedial students with a mean age of

14.6 in a mainstreamed social studies class. Findings revealed "students who had access to the hypermedia study guides exhibited better information retention than students who used alternative support and did not use the hypermedia study guides" (p. 410). Higgins, Boone and Lovitt conclude from this study " that for students with learning disabilities and remedial students, a hypermedia study guide is a viable educational tool that leads to retention of factual information" (p. 411).

A three year school-based research study conducted by Boone and Higgins (1993) investigated the use of hypermedia support and basal readers. Findings revealed that students who achieved the most significant reading gains were lower achieving students in the experimental group for the first and second grades. This study confirmed the effectiveness and use of multimedia/hypermedia software for special education students participating in the general education curriculum.

Multimedia

Daiute and Morse's (1994) research involving the use of multimedia writing tools for students with disabilities concludes that students can experience writing success through the power of multimedia. Such technology enables students to ground their writing assignments in their own world and personal experiences. Multimedia was found to be one way teachers can help children connect their specific perspectives and ways of expressing themselves to a common curriculum (Daiute & Morse, 1994).

Teachers are encouraged to allow students to guide the course of their learning to the extent that students are able. Through implementation of Anchored Instruction in the classroom, a need develops to change the culture of the classroom. The Cognition and Technology Group (CTGV) have identified three major challenges. First, the teacher role needs to change from provider of information to coach or mentor. Second, teachers need to know when students are in need of guidance versus when students are struggling in a constructivist way with a problem or issue. Third, an uncertainty exists about how to provide such guidance without being overly directive (The Cognition and Technology Group at Vanderbilt, 1993).

Electronic Performance Support Systems (EPSS)

"Hypertext, hypermedia, multimedia, databases, and expert systems are all the various technologies that can be used to support an electronic performance support system" (Brown, 1996, p. 5-2). The literature defines the field of electronic performance support systems (EPSS) as one of the most dynamic areas in human performance systems (Hudzina, Rowley & Wager, 1996).

Until recently, in the area of special education, similar supportive systems have been implemented on the administrative rather than the classroom or student level (Geldern, Ferrara, Parry & Rude, 1991; Prater & Ferrara, 1990). Currently, a new generation of EPSS is being developed to provide widely accessible computer-based literacy instruction for students with disabilities. One example of this new generation is the LiteracyAccess website and its Literacy Explorer© (the EPSS component). This website is an innovative effort in the field of special education to provide parents, advocates, instructional aides, and teachers a comprehensive means to teaching literacy skills to students with disabilities. Such systems provide one-on-one Internet-based literacy sessions by providing a simultaneous supportive environment for literacy learning (Bannan-Ritland, Egerton, Page & Behrmann, 2000).

Assistive Technology

The main goal of human performance technologies such as EPSS, is to provide the learner with whatever is necessary to ensure performance and learning success with a minimum of support from other people (Gery, 1995; Laffey, 1997). This goal parallels with the global goal of assistive technologies. Lewis (1998) "broadly conceptualizes assistive technologies as any technology with the potential to enhance the performance of persons with disabilities" (p. 16).

Currently, our nation provides special education and related services to 5.3 million children with almost half of these children being identified as having a specific learning disability (Herr, 1999). Assistive technologies are beginning to impact the field of special education. Such impact can be attributed to advancements in technology and federal mandates that have empowered individuals with disabilities and their advocates.

Smith and Jones (1999) note that within a very short period of time diverse and potentially more powerful devices and applications have emerged. Such available technologies include: text-to-speech and voice recognition software, hand-held scanning devices, touch-free switches, touch windows, voice input devices, and digitized speech. This recent increasing range and efficacy of assistive technologies is making it possible to address the diverse needs of individuals with disabilities at all levels including low incidence and high incidence disabilities (Smith & Jones, 1999).

Federal mandates recognizing the importance of technology include the Technology-Related Assistance to Individuals with Disabilities Act Amendments of 1994 (P.L. 103-218), also referred to as the Tech Act; the Individuals with Disabilities Education Act Amendments of 1997 (P.L. 105-17), also referred to IDEA; and the Americans with Disabilities Act (P.L. 101-336), also referred to as ADA. Figure 1 illustrates the federal laws that support the use of technology for individuals with disabilities.

Such mandates began to build the infrastructure needed to enhance the integration of assistive technology devices and/or services into the lives of individuals with disabilities (Smith & Jones, 1999). Through such mandates, states are required to identify and implement procedures for assistive technology. Smith and Jones (1999), in reference to the Tech Act, state:

> By mandating that states take specific actions, the Tech Act sought to improve access and timely acquisition of AT devices and services for individuals with disabilities. As AT devices and services become further accessible, IEP teams will have additional tools to consider in meeting the appropriate education needs of the student with disabilities (p. 254).

The Individuals with Disabilities Education Act Amendments of 1997 (IDEA) (PL 105-17), mandate that the Individual Education Plan (IEP) team must consider the assistive technology needs of every child receiving special education services. Wehmeyer (1999) in reference to IDEA, states:

> Public school agencies must consider the assistive technology needs of every student receiving special education services, and not just students for whom the IEP team determines assistive devices or services are necessary. It is likely that this requirement will heighten the attention to and, hopefully, increase access to assistive technology for children and youth with disabilities (p. 49).

According to Lewis (1998), assistive technology has two fundamental purposes. "First, it can augment an individual's strengths so that his or her abilities counterbalance the efforts of any disabilities. Second, technology can provide an alternate mode of performing a task so that disabilities are compensated for or bypassed entirely" (p. 17).

Bryant and Bryant (1998) stressed that "the selection of assistive technology for an individual must be guided by the setting-specific demands, the capabilities a person must possess to use the device, and the individual's functional limitations that will be bypassed by using the device" (p. 47). Higgins and Boone (1996) emphasized that the key to effective technology is the right match between the technological tools, problems, and implementation.

The use of assistive technology can provide students the opportunity to read and write. The research literature in the area of assistive technology is limited. Traditionally, assistive technology is viewed as tools for individuals with severe or multidimensional disabilities. A survey conducted by McGregor and Pachuski (1996) in the state of Pennsylvania revealed " despite the greater incidence of students with learning disabilities and mental retardation within

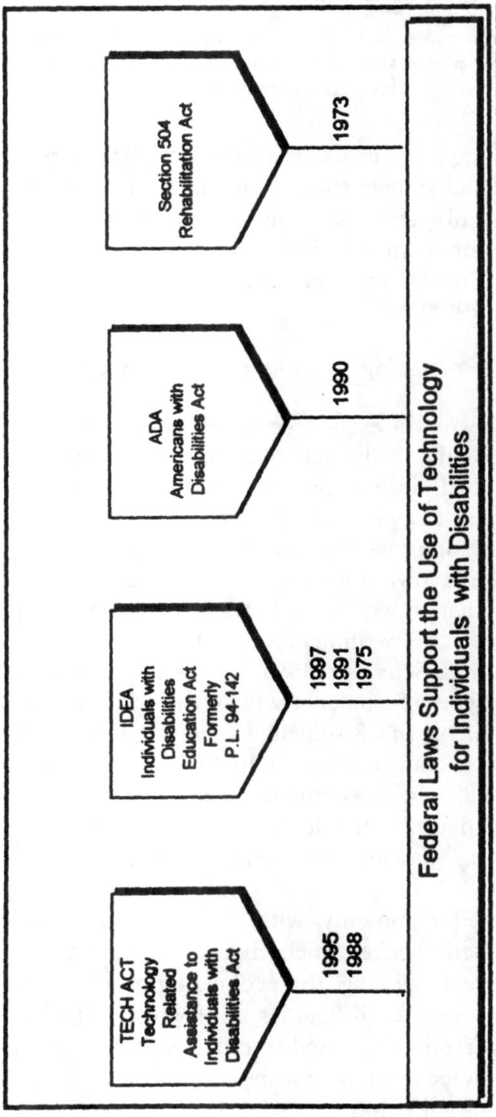

Fig. 1. Federal Laws That Support the Use of Technology.

their state; teachers associated assistive technology primarily with students with physical or multiple disabilities" (p. 12). McGregor and Pachuski state:

> If assistive technology comes to be viewed as only applicable to students with physical disabilities, it will violate the spirit of the federal law (i.e. the Tech Act) that intended to make these services more readily available to all people with disabilities who can benefit from such support (p. 12).

Today, more and more researchers are investigating the importance of assistive technology for all individuals (those with mild and severe disabilities) and attempting to not only understand its effectiveness and best practices for specific disability populations in implementing such technology, but also to validate the use of technology in schools, at work, in the home, and in recreation (Bryant, 1998).

Word Processing, Spelling and Grammar Checkers

Uses of word processors are quite common in our schools. MacArthur and Graham (1987) found that without strategy instruction, simply having access to a word processor had little impact on students' writing. A meta-analysis conducted by Bangert-Drowns (1993) concluded that the quantity and quality of students' writing improved but did not impact students' writing style/conventions or attitude toward writing. Further research validated that writing instruction in combination with word processing can significantly increase the amount and quality of writing for students with learning disabilities (MacArthur, Graham & Schwartz, 1991; Graham & MacArthur, 1988).

In a study involving 15 students with learning disabilities, students with severe spelling problems ($n = 8$ students) averaged 30% to 60% errors in their handwritten stories as compared to 6% to 17% errors in their word-processed stories (Outhred, 1989). In this particular study, word processing was found to be the most beneficial to students who had physical difficulties with handwriting and for students with severe problems in spelling (Outhred, 1989).

Another tool used commonly with students who have severe spelling difficulties is the spell checker. Spell checkers change the writer's task from producing the correct spelling to recognizing it (MacArthur, 1996). For students who have severe difficulties in spelling, spelling checkers may improve spelling, motivation, and build vocabulary (MacArthur, 1999). MacArthur (1996) noted even though spelling checkers are useful, for students with severe spelling difficulties, spell checkers have two limitations: (a) spelling checkers often are unable to suggest the correct spelling for severe misspellings, and (b) students are unable to identify the correct spelling in the suggested list of words. MacArthur and fellow researchers (Graham, DeLaPaz) have spent numerous efforts in investigating the effects of spell checkers on writing samples of students with learning disabilities. MacArthur, Haynes and Graham (1994) as cited in MacArthur, 1996, conducted a study that involved 26 middle school students with moderate to severe spelling problems and the use of spell checkers. Findings indicated that students were able to correct 82% of the errors with correct suggestions and 23% of the errors when the correct suggestions were not offered.

Word Prediction

Word prediction software can benefit students with physical/motor limitations and students with severe spelling problems. Lewis, Graves, Ashton and Kieley (1998) describe word prediction software as:

> When the first letter of a word is typed, the program attempts to predict the word being entered and lists several alternatives. If the correct word is displayed on the screen, the student simply selects it .... If the correct word is not displayed, the student types the second letter to see a second array of choices. This process continues until the correct word is available for selection (p. 96).

MacAuthur (1999) states that it appears that word prediction can make a substantial difference for those individuals with severe writing problems that interfere with the readability of their writing. The effectiveness of word prediction depends on the motivation and skill of the user, the word prediction software used, and the match between the software and the writing task at hand.

Multiple single-subject research studies conducted by MacArthur (1998) revealed various results in the use of word prediction programs. In one study, five students were asked to write dialogue journals to their teachers. Results revealed that word prediction software increased the proportion of correctly spelled words from a range of 42% to 75% to a range of 90%–100%.

In a study conducted by Lewis, Graves, Ashton and Kieley (1998), handwriting, word processing, alternative keyboards, and word prediction (with and without speech) were compared for writing speed, accuracy, and quality of students with and without learning disabilities. Results revealed that the speed of the handwriting group remained approximately the same from pretest to posttest. All other groups showed a decrease in speed. A closer look found that word prediction with speech had the most dramatic decrease in speed and in contrast, the word prediction group without speech seemed to sustain the least severe decrease in writing speed.

For the area of accuracy, Lewis et al. (1998) found that "students with learning disabilities made more errors at pretest than posttest. In the pretest-writing sample, students made an average of 37.1 errors per 100 words; at posttest, the average number of errors decreased to 29.4" (p. 102). Lewis et al. also emphasized "overall, students with learning disabilities had more writing errors than their general education peers" (p. 102).

Analytic writing scales were used to evaluate the quality of students' pretest and posttest first draft writing samples. Results indicated that the word prediction group had the highest quality scores and was superior to the word processing, keyboarding, and alternate keyboard groups.

Planning, Outlining and Organizing

Research has validated that the writing process should be taught to students with disabilities in an explicit manner. Troia, Graham and Harris (1999) emphasize:

> Such instruction should occur in an environment in which students' skills in self-regulation can prosper and grow allowing students to choose their own writing topics, assigning topics that are designed to serve a real purpose, encouraging students to share an environment that is supportive, pleasant, and non-threatening (p. 249).

Graphic organizers, structured overviews, and concept maps have been used to review and present instructional materials and concepts. Such instructional aids have their roots in Ausubel's advance organizer (Robinson, 1998). Ausubel's rationale for using advanced organizers was to provide students with a meaningful conceptual framework, in terms they understand, to help them organize newly acquired information (Robinson, 1998).

Graphic organizers have been used in a variety of content areas over the past two decades. Research demonstrates that such organizers are effective in increasing students' knowledge and comprehension of subject matter (Doyle, 1999; Penn, Shelley & Zaininger, 1998; Quist, 1995). Students can rely on graphic organizers to assist them in keeping focus on a topic and in creating clear and concise relationships between different types of information. Students and teachers use organizing tools, such as graphic organizers, to deepen the understanding of concepts and to enhance higher order thinking skills.

Dictation and Voice Recognition

The advancement and improvement of voice recognition technology is changing rapidly. De La Paz (1999) states "in anticipation to such technological advancements, a small but growing group of researchers has conducted research during the past 10 years to determine how this technology might best be used with persons with learning and writing problems" (p. 174).

Voice recognition software has the potential to allow individuals with disabilities to focus on high-level planning and organizing of content generation rather than on mechanics and physical writing (De La Paz, 1999). Only a few research studies have examined the use of voice recognition/ dictation to improve writing for students with disabilities. Research conducted by MacArthur and Graham (1987) examined the preliminary effectiveness of dictation for students with learning disabilities. In this study, fifth and sixth graders were asked to write a story from pictures. Results indicated that dictated stories were of higher quality, longer in length, and had fewer errors than when writing with pen or pencil or using a word processor.

A follow-up study conducted by Graham (1990) involved fourth and sixth graders; writing opinion essays one with normal dictation, one with slow dictation (composition was dictated and transcribed at the same speed that an earlier paper was written) and one hand written. Findings indicated that normally dictated essays were of higher quality than handwritten essays and were generated seven times faster for fourth graders and five times faster for fifth graders. One additional finding revealed that the mechanical demands of writing were disruptive for students.

Wetzel's (1996) research study examined the use of voice recognition software with three students in the intermediate grades (4th, 5th, 6th). He found that individual differences in verbal fluency and task selection influenced the usefulness of voice recognition. Students experienced frustration with articulating clear utterances and trying to correct recognition errors.

De La Paz (1999) highlighted the following recommendations for students using voice recognition systems: (a) planning is a central element to effective use of voice recognition systems; (b) scaffolding is needed as students learn the technology; (c) voice recognition and speech output could be helpful in the editing and revising process, and (d) a continual need exists for writing conventions such as mechanical and stylistic mechanics. I believe a fifth recommendation could be added to this list; (e) continual updating of voice recognition software to obtain the maximum benefits of a rapidly growing technology.

Supported Text and Text Readers

There is limited research in the area of supported text and text readers. A study conducted by Montali and Lewandowski (1996) revealed that poor readers benefited greatly from bimodal reading (the use of visual and auditory stimuli) when using the computer to read passages of text. Furthermore, " bimodal presentation of passages brought comprehension for less skilled readers up to a level comparable to the level of comprehension seen in the average group when they read the passages by themselves" (p. 276). Perhaps one explanation of such findings for a bimodal benefit is that by eliminating the struggle with word recognition, more focus could be placed on understanding the meaning of the passage (Montali & Lewandowski, 1996). In addition, a questionnaire revealed that poor readers prefer bimodal presentation of the text.

A study conducted by Wepner (1990) revealed that computer-reading software that provided the option for the reader to select the mode for reading each screen (word-by-word or by phrases) demonstrated to be effective with below average readers. Below average readers also selected the text to speech option before reading, thus listening intently until the story's end. "Overall, students saw the computer as an opportunity to explore the text without fear of failure, thereby developing a sense of self-control, inner power, and autonomy with their learning" (Wepner, 1987 as cited in Wepner, 1990).

Research also supports that assistive technology, used in combination with instructional strategies, can yield benefits for students with disabilities (MacArthur, 1999; Bryant & Bryant, 1998; Lewis, 1998). In addition, research reveals that students with disabilities do not necessarily take advantage of features of software unless prompted to do so.

Lewis (1998) addressed the use of emergent technologies for the future. The new networking capabilities, such as the Internet and the World Wide Web, have the potential for increasing the access of persons with disabilities to tremendously rich worldwide information resources.

Educational Research of the Internet

Intervention research involving the use of the Internet in K-12 classrooms provides insight regarding the impact of emerging technologies on the education and the learning process of students with disabilities. Such research enhances the opportunity to: (a) examine innovative and diverse ways the Internet has been implemented; (b) analyze the specific content areas that have initiated Internet usage, and (c) synthesize what the efficacy data suggest regarding the Internet as an educational tool.

Impact on Teacher Views and Uses

Teachers are the implementers of new teaching and learning tools in the classroom. In order for any tool to be used successfully the teacher must feel comfortable in implementing it into their curriculum. Two studies examined teachers' perceptions and use of the Internet as an educational tool.

In a case study, conducted by Alagbe and Lemlech (1998), patterns of use, effects on roles in relationships, and problems and preparation needs posed by the use of the Internet in four middle school classrooms were examined and described. School administrators, peers, and self-reports identified the four teachers selected to participate in this study as users of the Internet in their classroom instruction. An initial interview disclosed the intent of the unit in which they were going to implement the Internet. Data and information on teacher and student behaviors, as well as the role technology played in the educational process, was gathered through classroom observations. Post interviews with students and teachers focused on how students and teachers felt about the instructional unit and the role of technology. Research findings revealed that integrating the Internet into the classroom began to change the classroom community and roles of the teacher and student. Students needed to ask probing questions and teachers needed to guide the students through the learning process. Students communicated their findings instead of teachers telling students what is significant. Students needed to be empowered to explore and experiment and teachers' goals needed to be flexible. More teaching strategies were needed to arouse interest and stimulate discussion and less time needed to be devoted to presenting information (Algabe & Lemlech, 1998).

Slekar (1997) investigated two reciprocal influences of the Internet – the influence that the teacher's philosophy of teaching history had on choosing resources from the Internet, and the influence that the Internet had on this teacher's philosophy of teaching history. It was discovered that the teacher began to demonstrate a shift in teaching philosophy from teaching fact-based objectivist history to understanding the importance of the interpretive nature of history.

Impact on Learner Views and Uses

Understanding how students are using the Internet and what organizational processes and strategies they may employ while utilizing the Internet could provide insight to curriculum and instruction planning, along with providing information necessary in designing instructional supports needed for student success. Three research studies employed a case study approach (Craig, 1998; Lyons, Hoffman, Krajcik & Soloway, 1997; Oliver et al., 1997) to examine the impact of the Internet on the students.

Craig (1998) incorporated the use of the Internet as a tool in the process of inquiry for five fifth grade students. She examined field notes, video and audiotapes of "computer talk" (talk occurring while using the computer), work patterns, and the inquiry process. Findings revealed that students were not accustomed to having the freedom to select and pursue a research topic.

Lyons et al. (1997) completed a similar investigation (i.e. used video and audio tapes to record student interactions) examining online work patterns of students in sixth and ninth grade science classrooms (two pairs of middle school students and two pairs of high school students). Two separate instructional units (one for middle school students and one for the high school students) were distributed to the students. These units contained information on "Places to Start" and necessary background information that provided structure to the assignment. Lyons et al. (1997) research findings suggested that students need quite a bit of support in their work and flexible timeframes in order to be successful online.

A case study conducted by Oliver et al. (1997) revealed that fifth grade students displayed an ability to work independently of the teacher, requiring only minimal assistance while conducting "Web" searches. Students were also able to incorporate the information found from their Internet search into a final written project.

Two other research studies involving the examination of learner views and the use of the Internet employed quantitative research methodology (Martin, 1998; CAST, 1996). Martin (1998) investigated a sample of 14 pairs of fourth grade students. Simple measures for outcomes such as the total time to complete the assignment, the number of missteps in the Internet search process, the number of times teacher assistance was needed, and the initiation and involvement ratio for each pair of students were analyzed. Martin explored specifically the work patterns and learning strategies between female and male students. Although not statistically significant, some interesting patterns were revealed. For example, females tended to be more efficient than males, making 43% fewer mistakes. Overall, female pairs required less teacher intervention than male pairs. In addition, research findings revealed that male pairs collaborated less than female pairs; an average of only 18.2% of total events, compared to 49.2% for the female pairs.

The Center for Applied Special Technology (CAST) completed a research study that involved seven major U.S. cities during the 1995–1996 school year. A total sample of 500–600 students (14 control and 14 experimental classes) was used. The goals of this study were to measure the effects of online use on student learning, including information processing, communication, and presentation skills. Students in the fourth and sixth grades were asked to explore civil right issues, pose questions and select topics for a project. The effects of online use on student performance and attitudes displayed statistically significant between the control group (no Internet) and the experimental group (use of the Internet) in four of the learning measures: (a) effectiveness of presentation and presentation of the full picture, (b) effectiveness of bringing together different points of views and (c) completeness of projects. Statistical *t*-tests displayed significant increases in the experimental group's report of using computers to help with basic skills, gather information, organize and present information, and do multimedia projects.

In an ethnographic study conducted by Christie (1998), approximately 30 students (nine years of age) attended technology workshops. The 30 children spent between 35 and 105 hours interfacing with technology outside the regular school setting. Gender and racial demographics of the participants were 59% girls, 41% boys, 69% White, 17% Hispanic, 7% African American, and 7% Asian. Three profiles on students described the enthusiasm of learning through the Internet. Children took ownership in their research because the children had the freedom of materials (sites on the Internet) and the selection of topics. Student range of ability was reflected in how the students approached their research. Christie stated that students gravitated toward learning materials they understood and could utilize.

Impact on Learning Activities

The Internet is a vast resource of unleashed power of information that could be used as a vital educational tool. The use of new technologies such as the World Wide Web requires teachers to provide innovative learning activities to promote the use of the Internet and innovative ways for students to learn in today's classrooms.

Cunningham, Kent and Muir (1998) designed an activity that would provide collaboration in a meaningful context and developed the CyberOlympics. The cyber games took place during 1996, the year of the Atlanta Olympic Games. While the Olympics were taking place many schools took the opportunity to run mini sporting events within their own schools. The CyberOlympics encouraged students to share training tips, quiz for understanding, and exchange other interesting information. The CyberOlympic World Wide Web pages provided a central information source for all those interested in the Games and allowed every school to have access to any data collected.

A questionnaire was sent to all participants and organizers of local CyberOlympics events. A total of 36 responses were received. Findings revealed that most of the schools involved used the CyberOlympics theme in a variety of cross-curricular topics (i.e. language, P.E., geography, history, home economics, and drama). Twenty-nine percent of the respondents said they had contacted all the participating schools at least once, and the remaining 71% had at least contacted some of the participating schools. Ninety-six percent thought that these contacts would continue after the CyberOlympic games. The Scottish and American schools have exchanged samples of pupil's work and have begun work on another collaborative project together.

Another innovative Internet project was conducted by Ewing, Dowling and Coutts (1997); who were co-directors of Superhighway Teams Across Rural Schools (STARS); a project led by Northern College within the U.K. Education Departments' Superhighways Initiative (EDSI). The STARS project was set up to investigate how the superhighway might bring new learning opportunities to pupils in remote rural schools. The overall aim of the project was to demonstrate how learning for students and professional development for teachers can be enhanced by the use of multimedia resources made available over existing and emerging communications network.

A combination of quantitative and qualitative approaches to evaluating the STARS project was used. Eighteen schools participated in the project. Students were divided into responsibility teams similar to the Star Trek theme and asked to complete problem-solving tasks. An extensive analysis of the project took place after one year. Through teacher and student questionnaires and interviews; the overall conclusion of the STARS project was that the use of the computer was seen as an additional layer in the learning experience, contributing to the overall richness of classroom learning (Ewing et al., 1997).

A third innovative learning project involves The Learning through Collaborative Visualization (CoVis) Project. This project attempts to transform science learning to better resemble the authentic practice of science. The CoVis project is comprised of a team of collaborating researchers communicating with 40 schools around the country. Participating students study atmospheric and environmental sciences through inquiry-based activities. The CoVis Project provides students with a range of collaboration and communication tools. These include: desktop video teleconferencing; shared software environments for remote, real-time collaboration; access to the resources of the Internet; a multimedia scientist's notebook; and scientific visualization software. The project seeks to understand how science education could take broad advantage of these capabilities; providing motivating experiences for students and teachers with contemporary science tools and topics (Lento, O'Neill & Gomez, 1998).

A case study involving two of the schools participating in the CoVis project revealed preliminary results that CoVis pedagogy and technology has changed patterns of classroom activity (D'Amico, Gomez & McGee, 1998). Through this preliminary study, it became obvious that teachers and others will have to make the CoVis activities a part of their assessment regimens.

In summary of the current Internet research findings and synthesis presented here, five major research hypotheses can be drawn from the data: (a) gender differences exist in the work patterns and student approaches to using the Internet (Martin, 1998; Christie, 1998; Craig, 1998); (b) writing skills and quality of writing projects have improved through the use of the Internet (CAST, 1996; Chang et al., 1998); (c) authentic learning opportunities and the availability of a variety of learning resources provide richness to the learning experience (Alagbe & Lemlech, 1998; Cunningham et al., 1997; Ewing et al., 1997; Lento et al., 1998); (d) through the use of the Internet, learning becomes more student-centered (CAST, 1996; Lyons et al., 1997; Christie, 1998; Craig, 1998); and, (e) overall, the use of the Internet increases student enthusiasm (Lyons et al., 1997; CAST, 1996; Craig, 1998; Cunningham et al., 1997; Lento et al., 1998).

Current research demonstrates that the benefits of using technology in the educational process include learner motivation, ownership, empowerment, and control. Technology provides an interactive opportunity for learner engagement in the learning process. By identifying and gaining a better understanding of characteristics of students with reading and writing disabilities and the role of technology in these processes, interventions can be designed and implemented to assist and overcome literacy difficulties and barriers.

MERGING TECHNOLOGIES

A recent qualitative investigation of parent/child dyads and their use of assistive technologies in a literacy experience on the Internet provided a risk-free learning environment for parent/child dyads to explore, implement and merge various technologies to build literacy skills (Jeffs, 2000). The purpose of this study was to: (a) identify the characteristics of parent/child dyads working together in literacy skill development; (b) depict the interactions of dyad members; (c) investigate the impact of technology (Internet, EPSS, and any Assistive Technology) on attitudes of the participants of a literacy process, and (d) reveal areas of support needed by parents to facilitate literacy learning. This study addressed the following research questions and their subcomponents.

(1) What were the characteristics of dyad members interacting with technology to build literacy skills?
   (a) What were the individual needs and attitudes in using technology in the literacy process?
   (b) How did each member of the literacy dyad use technology in the literacy process?
   (c) What were the advantages and disadvantages of using technology in a literacy experience?

(2) What were the attitudes of dyad members toward literacy skill development?
   (a) How did dyads describe themselves in having specific competencies in reading and writing?
   (b) Did technology change pre-existing motivation, writing production quality, and personal comments about the literacy process?
(3) What types of interactions were characteristic of dyad members?
   (a) How did each member of the dyad define his or her role?
   (b) How did technology impact these roles?
(4) How much planning and support was necessary in order for parents to teach literacy skills to their child?

OVERVIEW OF FINDINGS

During this research, data was collected on eight parent/child dyads through observations, a researcher log, dyad journals, videotapes, writing samples, and pre/post interviews during a six-week period. In this study, participants attended six consecutive Saturdays ($2\frac{1}{2}$ hours per session) to work on the computer and explore the Internet and available assistive technologies to develop literacy skills. Initially, five of the dyads had strong positive feelings about using the Internet and assistive technology, two of the dyads wanted to explore the technologies to see if it would be beneficial, and one dyad had very negative feelings about using technology for learning. By the end of the study, all dyads had strong positive feelings about using the Internet and assistive technology for exploring and developing literacy skills.

Current research suggests that the benefits of using technology in the literacy process include learner motivation, ownership, empowerment, and control. Using new technologies provides an interactive opportunity for learner engagement in the learning process. Thus, understanding the learner's instructional needs and characteristics enables the planning and design of effective instruction and learning environments.

Dyads Interacting with Technology to Build Literacy Skills

Behrmann (1998) stated that it is important to realize that the potential range of assistive technology devices is broad and should be taken in consideration for each individual. This was particularly true in this study. Participants represented a variety of disability challenges related to the wide spectrum of disability types. Participants included children with learning disabilities,

Down's Syndrome, and multiple disabilities (one involving a physical disability) and a sensory disability.

All parent dyad members had limited use and exposure to the computer and had minimal computer skills prior to this study. Three of the dyad members experienced difficulty with computer access; therefore alternative devices were used (i.e. a trackball, an onscreen keyboard (Discover: Kenx), and a 21-inch monitor).

Child dyad members were in grades four through six and were reported reading at least two grades below their current grade level. Through the examination of previous and current writing samples, observations, videotaping, and participant journals; typical writing difficulties for child dyad members that were revealed that included poor spelling, lack of details, and difficulty with organizing and expressing thoughts or ideas for writing. Typical reading difficulties experienced by child dyad members included word recognition of sight vocabulary and slow and laborious reading. Before participating in this research study (and using the Internet and available assistive technologies), child dyad members would avoid reading and writing tasks.

The actual physical task of writing was difficult for three of the child dyad members, and virtually impossible for two of the child dyad members, due to cognitive or physical limitations. Therefore, technology provided the outlet for written expression. During this research study, each dyad explored and discovered their own learning style and technological needs.

Higgins and Boone (1996) emphasized that the key to effective technology is the right match between the technological tool and the problem. During this study, child members explored available assistive technologies and could immediately decide if a specific piece of technology would, or would not, work for them. Parent dyad members saw the importance of their child making such a decision. They knew that if their child did not like using the technology for any reason, it would not be used. Making the right match was important to each member of the dyad. Parents in this study also stated that this exploration and learning of available technologies was not as complicated as they thought or as some school systems led them to believe.

Triangulation of data from current writing samples, post interviews, and observations revealed the positive impact of the Internet and other technologies, such as word processors (Microsoft Word), presentation software (Microsoft PowerPoint), graphic-based software (Storybook Weaver), and assistive technologies (Write Outloud, textHELP, Co:Writer, Dragon Natural Speaking Preferred) on the quantity and quality of the child dyad member's writing. Students' attitudes toward their writing products also turned positive and all child dyad members felt that their writing had improved through the use of the Internet and these technologies.

In this research study, dyads working together with informal writing instruction and the use of the Internet, word processors, graphic-based software, and presentation software. Data suggested that child dyad members' writing samples improved in both quantity and quality throughout the study.

In comparison, a meta-analysis conducted Bangert-Drowns (1993) found that when students used word processors the quantity and quality of students' writing improved but did not impact students' writing style/ conventions or attitude toward writing. Perhaps one difference between the meta-analysis findings and this research study's findings could be the use of more specific assistive technologies to assist in areas of weakness and the use of the Internet, graphic-based software, and presentation software, that provided an exciting and motivating production media for writing.

Another finding revealed in this study was that all the child dyad members preferred to use the text to speech software to read aloud the Internet or their completed writing passages. Within such software, text is spoken and highlighted on the computer screen. They used the text to speech software with speed, accuracy, and confidence. They would listen and watch intently to the entire text from beginning to end.

Montali and Lewandowski (1996) revealed similar findings. They found that poor readers preferred bimodal presentation of text (the use of visual and auditory stimuli) and they benefited greatly from bimodal reading when using the computer to read passages of text.

In addition, a study conducted by Wepner (1990) revealed that computer-reading software that provided the option for the reader to select the mode for reading each screen (word-by-word or by phrases) was effective with below average readers. Wepner explained that students viewed the computer as an opportunity to explore text without fear of failure, thereby, developing a sense of self-control, inner power, and autonomy with their learning.

A similar experience was found in this study. Child dyad members who normally avoided text and did not feel confident in the area of reading and writing were experiencing confidence and enjoyed being able to decide when to use the text to speech software in the literacy process. All dyads reported that the use of the Internet, in connection with text-to-speech software (i.e. textHELP), was beneficial in the areas of reading, writing motivation, and independence.

Other technologies that proved to be of great benefit to the dyads during their literacy experience included graphic organizers and graphic-based writing programs. Six of the eight dyads discovered that graphic organizers (Inspiration software, 1997) provided focus and a road map in the writing process. Those same six dyads found that graphic-based writing software programs (i.e. Storybook Weaver Deluxe, 1995) sparked imagination and creativity for writing.

In addition, all dyad members expressed that the advantages of technology in the literacy process far outweighed the disadvantages. All parent dyad members acknowledged the importance of immediate feedback that technology provided and believed that assistive technology should be a priority for all students regardless of their disability.

Attitudes of Dyad Members Toward Literacy Skill Development

Literacy instruction was of great importance to the parent dyad members in this study. They believed that an increase in literacy skills would assist their child in being successful in school and in life activities. This is parallel to Torgesen's (2000) statement that children who become adults with poor literacy skills are at a disadvantage in society. Each of the dyad parent members believed that if they did not seek out answers to helping their child to build essential literacy skills, then no one else would.

Child dyad members were extremely motivated in using the Internet and enjoyed making decisions in the topics selected and use of assistive technology in the literacy process. All dyad members strongly expressed that the use of the Internet and assistive technology stimulated interest and promoted engagement. Dyad members had the opportunity to explore authentic learning opportunities and connect prior knowledge to new learning experiences through the use of the Internet. Within this study, dyad members felt that the Internet added richness, meaning, and enthusiasm to literacy through self-selected topics of interest. Cross dyad interactions and motivation were reflected to the extent that dyads enjoyed bringing in and sharing their own pets including a Bearded Dragon, Hermit Crabs and Cockatiel during the course of the research study.

Parents were surprised by the enthusiasm and engagement in reading and writing. Similar findings were found in 93% of the Internet studies discussed earlier in this chapter. As in similar Internet studies (Martin, 1998; Christie, 1998; Craig, 1998) the learning activities were student-centered and authentic learning took place. In this study, increased motivation in reading and writing had become evident. Similar findings were revealed in 86% of the Internet studies discussed in this chapter.

Parent and child members had become part of a cooperative team in researching, organizing, and producing the written product. The ability for child dyad members to discuss and share their writing provided a catalyst for editing and polishing their writing projects. Dyads showed evidence of pride and ownership in their reading and writing activities. Similarly, when examining the Internet studies discussed in this chapter, findings revealed that 64% of the studies reported on newly acquired ownership of the learning project.

Types of Interactions were Characteristic of Dyad Members

Initially, child dyad members were more exploratory and at ease with using the computer than parent dyad members. By the end of the study, all dyads were at ease in exploring and using new technologies. Dyad members shared expertise when engaged in the literacy process using technology.

Similar findings were found between this study and the early pilot studies conducted as part of the LiteracyAccess Online Project (U.S. Department of Education (1998) H327A980035). Such findings were that: (a) children were motivated and independently sought out information on the Internet, (b) parents were eager to assist their child in building literacy skills, (c) tension and power struggles did exist between dyad members; and (d) all dyads experienced success using assistive technology to overcome access and interface issues in using the computer and the Internet.

This study not only confirmed earlier findings but also provided a greater insight and understanding of the common characteristics, interactions, and use of technologies by parents and their children with disabilities. Dyad members used a combination of technologies to meet specific individual learning needs. For example, dyads employed a text-to-speech software program (textHelp) to assist in reading, a word prediction program (Co:Writer) to assist in spelling, and a graphic-based writing program (Storybook Weaver) to assist in writing and publishing.

Careful attention was also given to dyad interactions and the transformation and learning the dyads experienced while working with one another. Dyad members learned from each other and from the technology. They began to integrate technology to meet their individual needs. Customization of the learning task through technology enhanced opportunities for engagement and interaction to take place. Dyads began to explore and understand each other's sense of organization, details, agenda, and learning differences.

Ownership played upon the roles of the dyads. Both members had a vested interest in the writing product. Negotiations were difficult at times creating conflict within dyads. Overall, dyads eventually compromised on organization, details, and learning differences that they were just starting to understand.

Planning and Support Needed for Parents to Teach Literacy Skills

Parent dyad members enjoyed working with their child on the Internet and assisting them in the literacy process in this study. They also enjoyed planning and discussing reading and writing activities with their child. Parents commented that it was great to spend time together and this quality time seemed to have a positive effect not only on literacy skills and attitudes but also on their parent-child relationship. Parents saw the importance of immediate and positive feedback, no matter if given by technology or by the parent.

Research findings by Lyon et al. (1997) suggested that students need quite a bit of support in their work and flexible timeframes in order to be successful online. This was also true in this study. Dyad members worked as a team in the literacy process, utilizing each other's strengths and weaknesses. They found it was an advantage to discuss the reading and writing processes. Dyad members were in control of their own timelines, lesson structure, and writing products, and therefore, only felt pressure of these limitations by standards that they established for themselves.

It is obvious that, in order for the Internet and assistive technology to be used successfully in the schools, the learning environment must become more student-centered. Flexible timeframes and student-centered activities present new challenges to classroom management and behavioral strategies and techniques. Similar findings were revealed in a recent study investigating special education teachers' instructional applications of the Internet. Castellani (1999) found that there was a distinct struggle between behavioral intervention and academic learning. In the study, teachers believed that when students were engaged in self-directed activities, the learning experience was more meaningful and transferable (p. 200). Castellani reported that teachers in this study believed that "students needed freedom to explore materials for increased exposure to people, places, and things outside of their home or community experiences" (p. 201). Teachers within Castellani's study, expressed that an appropriate balance of student-oriented and teacher-directed activities was necessary to work with the Internet for students with learning disabilities and behavioral disorders. With the current high demands on the school curriculum (eg. the Standards of Learning tests) pressure may be too great to implement a new style of teaching.

All dyads spoke of their struggles with schools in providing technology as part of their child's educational program. Parents in this study voiced that there were too many inconsistencies in assistive technology services. Parent dyad members felt that school systems did not recognize individual needs when considering technology and expressed that school administration and teachers viewed assistive technology as technology needed only for individuals with physical disabilities.

This was a similar finding to McGregor and Pachuski's (1996) survey study that revealed that teachers associated assistive technology primarily with students with physical or multiple disabilities. As McGregor and Pachuski stated, if assistive technology is viewed in this narrow sense then it simply violates the spirit of the Federal Law (i.e. Tech Act, 1988). A need exists for teacher education programs to include preservice education on how to effectively integrate available assistive technology into the curriculum to meet the needs of the students.

The eagerness of parents to get involved and stay involved in literacy instruction has promising benefits for schools and parents to work together. Although this study exposed parent struggles in getting schools involved in accommodating student needs through technology, each parent felt that it would be beneficial to provide schools with resources and information of assistive technology and how it could be implemented in the literacy process.

During this research study, parents asked for specific literacy information, strategies and possible activities. As a participant-observer, I would provide very short one- or two-word strategies, to try. Parents were pleased with the suggestions, (i.e. let's brainstorm, imagine it differently, take a break, and test it out). Such tips and strategies were taken from the existing LiteracyAccess Online electronic performance support system (EPSS) of the Literacy Explorer. Although more research is needed, this could possibly indicate that parents might appreciate "the just in time" suggestions offered through the EPSS in the Literacy Explorer. The LiteracyAccess website and Literacy Explorer is an innovative effort in the field of special education to provide parents, advocates, instructional aides, and teachers a comprehensive means to teaching literacy skills to students with disabilities.

Specific feedback was solicited from the dyad members to state what a literacy website should contain in order for both dyad members to be successful in the literacy process. Dyad members wanted to see assistive technology features built right into the website. They suggested that the website should offer reading selections at varying ability levels and areas of interest. They also wanted multimedia and graphics to provide an interactive reading format. After completing this study, all parent dyad members believed that they and others could learn how to teach their child literacy skills through the Internet.

Technology provides a powerful and responsive environment that involves the learner in the learning process and gives consideration for individual learning styles, motivations, and goals (Cooper, 1993). The dyads experienced this powerful environment during this study. Parent dyad members were provided with basic resources and technological tools needed to make learning successful for their child's individual needs. This made the literacy experience more accessible, meaningful, and motivational for child dyad members. By better understanding the learner characteristics, interactions and attitudes, such resources and tools can be even more beneficial in the literacy development process.

Research conducted by CAST (see website, http://www.cast.org) emphasized the importance of universal design in order to meet the unique needs of learners. CAST defines the basic premise of universal design for learning as the inclusion of text alternatives to make text accessible and applicable to all users regardless of different backgrounds, learning styles, abilities, and disabilities. Universal design eliminates the premise of "one size fits all" learning. Innovative technologies are creating new opportunities in making information more accessible through customization, versatility, flexibility of printed text, electronic text on the World Wide Web, and the general academic curriculum.

IMPLICATIONS FOR RESEARCH AND PRACTICE

This study found that dyads were motivated and enthusiastic in using the Internet. Dyads voiced that graphics, video, and other multimedia features made literacy come alive. The Internet provided a global learning environment that enabled dyads to be actively engaged in the learning process through the exploration and manipulation of learning resources online with the use of assistive technologies.

This study found that the Internet provided meaningful, self-directed, and authentic instruction. Child members sought out information on the Internet and constructed their own meaning to information that they found. They took this information and created a writing product that had value and authenticity in their lives. Dyads not only practiced basic skills, but they also had the opportunity to apply higher-order thinking skills through the application of gathered information. Further research investigating instructional activities using the Internet for authentic learning may reveal common threads that could be beneficial across disciplines and ability levels.

This study found that dyads would customize their use of technologies to meet their specific needs and learning characteristics. In addition, by understanding the common characteristics, interactions, and attitudes of parent/child dyads and their use of such resources revealed in this study such customization and interventions could be designed and implemented to assist and overcome literacy difficulties and barriers. Thus in turn, researching developing technologies would open doors for all students to learn and benefit from new and innovative technologies.

In this research study, all dyads used assistive technology in conjunction with the Internet to explore reading and writing opportunities. The use of assistive technologies played a crucial role. Future research could compare the use of the Internet in the exploration of literacy activities with and without assistive technology. Such an investigation would explore if the Internet alone would provide the excitement, motivation, and engagement for students with varying exceptionalities or if assistive technology allowed the Internet to become an accessible educational tool for students with disabilities. Such a study would provide insight into accessibility, the exploration and development of literacy skills on the Internet, and online learning environments in general. Similarly, a study that compares assistive technology tools available through the Internet with stand-alone devices would be beneficial. The role of assistive technologies for students with disabilities would be further understood.

This study found that literacy dyads stayed extremely motivated and enthusiastic about reading and writing for the entire six weeks of the project. As a result, research is needed to examine the factor of time. Was the motivation and engagement in this current research a result of learning something new (novelty)? Would findings be similar to this study if given the opportunity to conduct the study over an entire academic year?

This study found that parent dyad members were frustrated by the lack of cooperation and partnerships with their child's school. Parents voiced that schools should be involved in, or at least informed of, the use of technology in the literacy process. Furthermore, research data from this study could assist in the building of partnerships between home and school through the development of innovative Internet-based programs at the local, state, and national levels.

U.S. Secretary of Education Richard Riley has challenged Americans for greater family, school, and community partnerships (i.e. Federal programs such as the Compact for Reading and Start Early, Finish Strong programs). Such programs could involve parent/child dyads working together at home using the information from the Internet as an extension to the school day. For example, dyad members could work together in practicing basic academic skills or researching specific content area topics through activities on the Internet, or writing reflections about special topics discussed in the classroom. By understanding dyad characteristics and interactions, the development and implementation of such activities could be successful for students with disabilities.

SUMMARY

> It takes time to figure out new technology – to discover the valuable new uses implicit in the technology itself" (Meyer & Rose, 1998, p. 7).

The voices of the dyads involved in this study revealed that learning is changing and even parents with minimal exposure to technology truly believe that in our fast-paced, information-driven society technology is a viable answer for their child. With the development and implementation of Internet2's capabilities to deliver faster and more efficient Internet service, there is little doubt in this researcher's mind that the Internet of tomorrow will consist of the multimedia-rich web sites as requested by the dyads in this study.

Careful customizations of such technologies as the Internet provide accessibility and opportunities for success in the literacy process. Galbreath (1999) states that the web provides us with the power of customization within itself. The web allows us to research topics of interest and implement online resources into our academic, professional, and personal lives. The Internet, in combination with careful matching assistive technologies to meet individual needs, takes customization to a higher level.

By examining such customization and use of technologies, instructional designers and other professionals designing learning curriculum can begin to implement preliminary assistive technology tools, using universal design for learning and making online learning truly accessible to all individuals. The capabilities of the Internet should provide endless possibilities for interactive learning. The fields of Instructional Technology and Special Education can only benefit from continued research in examining the Internet and other technologies as effective tools in creating new and innovative learning instruction and environments. The growth in development of technology has clearly surpassed the implementation of classroom strategies (Metheny, 1997). Special educators are accustomed to making curriculum adaptations; the time has come to train special educators to make technology-based curriculum/ learning adaptations.

REFERENCES

The Americans with Disabilities Act of 1990, Pub. L. No. 101–336 (1991).

Alagbe, A. F., & Lemlech, J. K. (1998). Middle school teachers' use of on-line communications. Paper presented at the Annual Meeting of the American Educational Research Association, San Diego, CA.

Baker, L., & Wigfield, A. (1999). Dimensions of children's motivation for reading and their relations to reading activity and reading achievement. *Reading Research Quarterly, 34*, 452–477.

Bannan-Ritland, B., Egerton, E., Page, J., & Behrmann, M. (2000). Literacy explorer: A support tool for novice reading facilitators. *Performance Improvement Journal*, *39*(6), 47–63.

Bangert-Drowns, R. L. (1993). The word processor as an instructional tool: A meta analysis of word processing in writing instruction. *Review of Educational Research*, *63*, 69–93.

Behrmann, M. (1998). Assistive technology for young children in special education. In: C. Dede (Ed.), *Leading with Technology* (pp. 73–96). Alexandria, VA: Association for Supervision and Curriculum Development Yearbook.

Boone, R., & Higgins, K. (1993). Hypermedia basal readers: Three years of school-based research. *Journal of Special Education Technology*, *12*, 86–106.

Brown, L. A. (1996). *Designing and developing electronic performance support system*. Boston, MA: Digital Press.

Bryant, B. R. (1998). Assistive technology: An introduction. *Journal of Learning Disabilities*, *31*, 2–3.

Bryant, D. P., & Bryant, B. R. (1998). Using assistive technology adaptations to include students with learning disabilities in cooperative learning activities. *Journal of Learning Disabilities*, *31*, 41–54.

Castellani, J. D., (1999). Technology and teacher education: A qualitative study exploring the instructional applications of the Internet in a setting for students with emotional and learning disabilities. Unpublished doctoral dissertation, George Mason University, Fairfax, Virginia.

Center for Applied Special Technology (CAST) (1996). The role of online communications in schools: A national study. (On-line). Available: http://www.cast.org (2000, Sept, 9).

Chang, H., Honey, M., Light, D., Moeller, B., & Ross, N. (1998). *The Union City story: Education reform and technology, students' performance on standardized tests*. New York: Education Development Center.

Christie, A. (1998). Using technology to enhance literacy in elementary schools. Paper presented at the National Educating Computing Conference, San Diego, CA.

Cognition and Technology Group at Vanderbilt (1993). Examining the cognitive challenges and pedagological opportunities of integrated media systems: Toward a research agenda. *Journal of Special Education Technology*, *12*, 118–124.

Cooper, P. A. (1993). Paradigm shifts in designed instruction: From behaviorism to cognitivism to constructivism. *Educational Technology*, *33*, 12–19.

Co:Writer [Computer Program] (1998). Volo, IL: Don Johnston.

Craig, D. V. (1998). Learner in charge: Student inquiry and technological literacy. Paper presented at the National Educating Computing Conference, San Diego, CA.

Cunningham, M., Kent, F., & Muir, D. (1998). The Cyber-Olympics – schools, sports, and the superhighway. *Computers in Education*, *30*, 61–65.

Daiute, C., & Morse, F. (1994). Access to knowledge and expression: Multimedia writing tools for students with diverse needs and strengths. *Journal of Special Education*, *12*, 221–253.

De La Paz, S. (1999). Composing via dictation and speech recognition systems: Compensatory technology for students with learning disabilities. *Learning Disabilities Quarterly*, *22*, 173–182.

Dillion, A., & Gabbard, R. (1998). Hypermedia as an educational technology: A review of the quantitative research literature on learner comprehension, control, and style. *Review of Educational Research*, *68*, 322–349.

Discover: Kenx [Computer Program]. (1998). Volo, IL: Don Johnston.

Doyle, C. S. (1999). The use of graphic organizers to improve comprehension of learning disabled students in social studies. (ERIC Reproduction Service Document No. ED 427 313)

Dragon Naturally Speaking Preferred, 4.0 [Computer Program] (1999). Newton, MA: Dragon Systems.

Duffy, T. M., & Cunningham, D. J. (1996). Constructivism: Implications for design and delivery of instruction. In: D. H. Jonassen (Ed.), *Handbook of Research for Educational Communications and Technology* (pp. 170–199). New York: Macmillan.

Englert, C. S., & Raphel, T. E. (1988). Constructing well-formed prose: Process, structure, and metacognitive knowledge. *Exceptional Children, 54*, 513–520.

Ertmer, P. A., & Newby, T. J. (1993). Behaviorism, cognitivism, constructivism: Comparing critical features from an instructional design perspective. *Performance Improvement Quarterly, 6*(4), 50–72.

Ewing, J., Dowling, J., & Coutts, N. (1997). STARS: Report on superhighway teams across rural schools project. (ERIC Document Reproduction Service No. 421 319)

Galbreath, J. (1999). Preparing the 21st century worker: The link between computer-based technology and future skill sets. *Educational Technology, 39*, 14–22.

Geldern, L., Ferrara, J. M., Parry, J. D., & Rude, H. (1991). Local validation of mandate consultant: An expert system for assuring compliance in the IEP process. *Journal of Special Education Technology, 11*, 113–120.

Gery, G. (1995). Attributes and behaviors of performance-centered systems. *Performance Quarterly, 8*, 47–93.

Graham, S. (1990). The role of production factors in learning disabled students' compositions. *Journal of Educational Psychology, 82*, 781–791.

Graham, S., & MacArthur, C. (1998). Improving learning disabled students' skills at revising essay produced on a word processor: Self-instructional strategy training. *Journal of Special Education, 22*, 133–152.

Graham, S., Schwartz, S. S., & MacArthur, C. A. (1993). Knowledge of writing and the composing process, attitude toward writing, and self-efficacy for students with and without learning disabilities. *Journal of Learning Disabilities, 26*, 237–249.

Graham S., & Weintraub, N. (1996). A review of handwriting research: Progress and prospects from 1980 to 1994. *Educational Psychology Review, 8*, 7–87.

Hannafin, M. J., Hannafin, K. M., Hooper, S. R., & Rieber, L. P. (1996). Research on and research with emerging technologies. In: D. H. Jonassen (Ed.), *Handbook of Research for Educational Communications and Technology* (pp. 378–395). New York: Macmillan.

Hannafin, M. J. (1992). Emerging technologies, ISD, and learning environments: Critical perspectives. *Educational Technology Research and Development, 40*, 49–63.

Herr, S. S. (1999). Special education law and children with reading and other disabilities. *Journal of Law & Education, 28*, 337–389.

Higgins, K., & Boone, R. (1996). Special series on technology: An introduction. *Journal of Learning Disabilities, 29*, 340–343.

Higgins, K., Boone, R., & Lovitt, T. C. (1996). Hypertext support for remedial students and students with learning disabilities. *Journal of Learning Disabilities, 29*, 402–412.

Hudzina, M., Rowley, K., & Wager, W. (1996). Electronic performance support technology: Defining the domain. *Performance Improvement Quarterly, 9*, 36–47.

Inspiration, 5.0 [Computer Program] (1997). Portland, OR: Inspiration.

Jeffs, T. L. (2000). Characteristics, interactions, and attitudes of parent/child dyads and their use of assistive technology in a literacy experience on the Internet. Unpublished doctoral dissertation, George Mason University, Fairfax, Virginia.

Jonassen, D. H., Dyer, D., Peters, K., Robinson, T., Harvey, D. King, M., & Loughner, P. (1997). In: B. H. Khan (Ed.), *Web-Based Instruction* (pp. 119–134). New Jersey: Educational Technology.

Laffey, J. (1997). Dynamism in electronic performance support systems. *Performance Improvement Quarterly, 10*, 183–198.

Lento, E. M., O'Neill, K., & Gomez, L. M. (1998). Integrating Internet services into school communities. In: C. Dede (Ed.), *Leading with Technology* (pp. 141–168). Alexandria, VA: Association for Supervision and Curriculum Development Yearbook.

Lewis, R. B. (1998). Assistive technology and learning disabilities: Today's realities and tomorrow's promises. *Journal of Learning Disabilities, 31*, 16–26.

Lewis, R. B., Graves, A. W., Ashton, T. M., & Kieley, C. L. (1998). Word processing tools for students with learning disabilities: A comparison of strategies to increase text entry speed. *Learning Disabilities Research & Practice, 13*, 95–108.

Lyons, D. J., Hoffman, J., Krajcik, J., & Soloway, E. (1997). An investigation of the use of the world wide web for on-line inquiry in a science classroom. Paper presented at the Annual Meeting of the National Association for Research in Science Teaching, Chicago, IL.

MacArthur, C. A. (1996). Using technology to enhance the writing processes of students with learning disabilities. *Journal of Learning Disabilities, 29*, 344–354.

MacArthur, C. A. (1998). Word processing with speech synthesis and word prediction: Effects on the dialogue journal writing of students with learning disabilities. *Learning Disabilities Quarterly, 21*, 151–166.

MacArthur, C. A. (1999). Overcoming barriers to writing: Computer support for basic writing skills. *Reading and Writing Quarterly, 15*, 169–192.

MacArthur, C., & Graham, S. (1987). Learning disabled students composing under three methods of text production: Handwriting, word processing, and dictation. *Journal of Special Education, 21*, 22–42.

MacArthur, C., Graham, S., & Schwartz, S. (1991). Knowledge of revision and revising behavior among learning disabled students. *Learning Disability Quarterly, 14*, 61–74.

Martin, S. (1998). Internet use in the classroom: The impact of gender. *Social Science Computer Review, 16*, 411–418.

Mastropieri, M. A., & Scruggs, T. E. (1997). Best practices in promoting reading comprehension in students with learning disabilities 1976 to 1996. *Remedial and Special Education, 18*, 197–213.

McAlister, K. M., Nelson, N. W., & Bahr, C. M. (1999). Perceptions of students with language and learning disabilities about writing process instruction. *Learning Disabilities Research & Practice, 14*, 159–172.

McGregor, G., & Pachuski, P. (1996). Assistive technology in schools: Are teachers ready, able, and supported? *Journal of Special Education Technology, 13*, 4–15.

Meyer, A., & Rose, D. H. (1998). *Learning to read in the computer age*. Cambridge, MA: Brookline.

Metheny, R. (1997). Curriculum adaptations: Improving student success. Closing the Gap (Online). Available: http://www.closingthegap.com (2000, September, 9).

Microsoft PowerPoint [Computer Program] (1999). Redmond, WA: Microsoft.

Microsoft Word [Computer Program] (1999). Redmond, WA: Microsoft.

Montali, J., & Lewandowski, L. (1996). Bi-modal reading: Benefits of a talking computer for average and less skilled readers. *Journal of Learning Disabilities, 29*, 271–279.

National Institute of Child Health and Human Development National Institutes of Health (1999). The NICHD research program in reading development, reading disorders and reading instruction. NICHD- Keys to Successful Learning Summit. (Online) Available: http://www.ncld.org/summit99/keys99-nichd.htm (2000, September, 9).

Netscape, 4.5 [Computer Program] (1999).MountainView, CA: Netscape.

Newcomer, P. L., Barenbaum, E. M., & Nodine, B. F. (1988). Comparison of the story production of LD, normal-achieving, and low achieving children under two modes of production. *Learning Disabilities Quarterly, 11*, 82–96.

Oliver, B. E., Malm, L. D., Malone, B. G., Nay, F. W., Saunders, N. G., & Thompson, J. C. (1997). Enhancing elementary curricula, through Internet technology. Paper presented at the Mid-Western Educational Research Association, Chicago, IL.

Outhred, L. (1989). Word processing: Its impact on children's writing. *Journal of Learning Quarterly, 22*, 262–264.

Penn, C., Shelley, S., & Zaininger, L. (1998). Enhancing transfer of learning among seventh graders. (ERIC Reproduction Service Document No. ED 422 117)

Prater, M. A., & Ferrara, J. M. (1990). Training educators to accurately classify learning disabled students using concept instruction and expert system technology. *Journal of Special Education Technology, 10*, 147–156.

Quist, S. (1995). The effect of using graphic organizers with learning disabled students to increase comprehension. New Jersey. (ERIC Reproduction Service Document No. ED 379 646)

Robinson, D. H. (1998). Graphic organizers as aids to text learning. *Reading Research and Instruction, 37*(2), 85–105.

Slekar, T. D. (1997, March). Technology's impact on one teacher's philosophy of teaching U.S. History. Paper presented at the Annual Meeting of the American Educational Research Association, Chicago, IL.

Smith, S. J., & Jones, E. D. (1999). The obligation to provide assistive technology: Enhancing general curriculum. *Journal of Law & Education, 28*, 247–265.

Storybook Weaver Deluxe [Computer Program]. 1995. Cambridge, MA: The Learning Company.

Swanson, H. L. (1999). Reading research for students with LD: A meta-analysis of intervention outcomes. *Journal of Learning Disabilities, 32*, 504–532.

Technology Related Assistance for Individuals with Disabilities Act of 1988 and Amendments (Catalogue No. 850, Senate Rep. No. 100–438). Washington, D.C.: U.S. Government Printing Office.

textHELP: Read & Write [Computer Program] (1999). Antrim, N. Ireland: textHELP Systems Ltd.

Torgesen, J. K. (2000). Individual differences in response to early interventions in reading: The lingering problem of treatment resisters. *Learning Disabilities Research & Practice, 15*, 55–64.

Troia, G. A., Graham, S., & Harris, K. R. (1999). Teaching students with learning disabilities to mindful plan when writing. *Exceptional Children, 65*, 235–252.

Wehmeyer, M. L. (1999). Assistive technology and students with mental retardation: Utilization and barriers. *Journal of Special Education Technology, 14*, 48–58.

Wepner, S. (1990). Computers, reading software, and at-risk eighth graders. *Journal of Reading, 34*(4), 264–268.

Wetzel, K. (1996). Speech-recognition computers: A written-communication tool for students with learning disabilities? *Journal of Learning Disabilities, 29*, 371–380.

Write Outloud [Computer Program] (1998). Volo, IL: Don Johnston.

U.S. Department of Education. (1998). LiteracyAccess Online.Steppingstones of technology innovation for students with disabilities program/phase 1 – development. CFDA 84.327A.

U.S. Department of Education (1999). Office of Educational Technology. (On-line). Available: http://www.ed.gov/Technology/ (2000, September, 9).

U.S. Secretary of Education, Richard W. Riley (1999). (On-line). Available: http://www.ed.gov/PressRelease/07–1999/readtools.html (2000, September, 9).

CRITICAL ISSUES IN SPECIAL EDUCATION TECHNOLOGY RESEARCH: WHAT DO WE KNOW? WHAT DO WE NEED TO KNOW?

Dave L. Edyburn

ABSTRACT

The potential of technology for individuals with disabilities has long been recognized by the special education community. However, the rate of marketplace developments has far outstripped the research base documenting the effectiveness of specific applications of assistive and instructional technology. The purpose of this chapter is to summarize critical issues associated with six areas of special education technology: accessibility, assistive technology, professional development, instructional technology, service delivery, and legal/policy issues. Within each topical area, key works defining the current knowledge base will be summarized along with an analysis of important questions that deserve additional research. The implications of these analyses for defining future leadership and research agendas are noted.

The fields of special education and rehabilitation have a long-standing interest in technology and the potential it holds for individuals with disabilities. Several detailed histories of the field of special education technology document the evolution of the discipline and highlight events and products that significantly

Technological Applications, Volume 15, pages 95–117.

ISBN: 0-7623-0815-x

impacted the use of technology by individuals with disabilities (Blackhurst, 1997; Blackhurst & Edyburn, 2000; Fein, 1996; Hannaford, 1993; Hauser & Malouf, 1996; Hobson, 1996; U.S. Department of Education, 2000). A careful reading of the historical record suggests two powerful forces, innovation and policy, have operated in tandem throughout the twentieth century to advance the development, availability, and use of technology by people with disabilities.

Innovation

Innovation has largely occurred as a result of enterprising individuals and organizations as they created technological devices to augment human abilities. For example, the Audiophone Bone Conduction Amplifier was invented in 1874 and by 1900 the first electrical amplifying devices for individuals with hearing impairments was available. As radios made their way into American homes in the 1920s, in 1928 the American Foundation for the Blind distributed radios to blind citizens to provide them with access to information that previously was only available in print formats. Similarly, as the recording industry emerged in the 1930s in response to Edison's invention of the phonograph, talking books for the blind were produced on long-playing phonograph records as early as 1935. Fast forwarding to computer era we can observe that only a few years following the introduction of the first Apple computer in 1977, the Adaptive Firmware Card (AFC) was invented. The AFC allowed individuals with disabilities to operate the computer using a single-switch or with an alternative keyboard. The AFC offered access to computer-based learning and productivity tools during the early days of the personal computing.

These few examples illustrate four historical patterns. First, the disability community has been quick to recognize the potential value of specific technological advances and capitalize on innovations by developing relevant applications for individuals with disabilities. Second, the lag between new technological advances and the availability of a specific disability application has been compressed during the twentieth century from over 20 years at the dawn of the century to only a few years at the end of the century. Third, while new technologies may initially be inaccessible to individuals with disabilities, modifications and innovative solutions routinely emerge to provide equal access. Finally, the disability community has demonstrated a consistent commitment to special education technology throughout the century.

Policy

The second powerful force influencing the advancement of special education technology in the twentieth century has been policy. Laws passed in the late nineteen century (i.e. 1879, P.L. 45-186 provided funds to the American Printing House for the Blind to produce Braille materials) and early 20th century (i.e. 1904, P.L. 58-171 promoted the circulation of reading matter among the blind) served to advance opportunities for individuals with disabilities who were perceived to be at a disadvantage in society. Over time, the disability community recognized the strategic value federal policy would serve for advancing the rights of individuals with disabilities.

Three specific legislative accomplishments benchmark the success of the disability rights policy agenda: 1973, P.L. 93-112 Rights of handicapped individuals in employment and educational institutions are guaranteed through Section 504 of the Rehabilitation Amendments; 1975, P.L. 94-142 Free appropriate public education and other procedural guarantees are mandated for all children with disabilities, and 1990, P.L. 101-336 passage of the Americans with Disabilities Act, landmark legislation defining legal protections for individuals with disabilities, requiring employers to make reasonable accommodations, and mandating accessibility to public buildings. The accomplishments associated with this legislative agenda provided a context for political activism that contributed significantly to policy advances concerning technology in special education.

In the modern era, one of the first public policy documents to draw attention to the potential of technology was *Technology and Handicapped People* (U.S. Congress, Office of Technology Assessment, 1982). Readers were introduced to specialized technology tools, how they served a specific individual, and the impact these devices had on their lives. The real-life stories served as powerful illustrations of the potential of technology for individuals with disabilities. Success stories were an essential component of persuasive arguments that were advanced on the premise that public investment in research and development in the area of technology and disability could reap significant dividends in the form of improved communication skills, expanded mobility and independence, as well as an increase in the number of individuals gainfully employed and contributing to the tax base. Other marker events include passage of the Technology Related Assistance for Individuals with Disabilities Act (1988) which created a mechanism for each state to develop a responsive system for acquiring and training residents in their state on assistive technology and the 1997 reauthorization of IDEA which requires each IEP team to consider assistive technology when planning the educational program of each student with a disability.

Critical Issues in Special Education Technology Research

The dual impact of policy and innovation throughout the twentieth century created an expectation that the availability of technology and its use by individuals with disabilities would yield tremendous benefits. This expectation was formalized in the 1997 reauthorization of the Individuals with Disabilities Education Act (Public Law 105-17) through a requirement that technology consideration must be a routine part of educational planning.

As the discipline of special education technology began to emerge during the 1980s and 1990s, research efforts became more notable. The organization of the Technology and Media (TAM) division of the Council for Exceptional Children in 1984 established the first membership organization where special educators with an interest in technology could affiliate. TAM is the publisher of one of the premier research journals for the discipline: *The Journal of Special Education Technology* (http://jset.unlv.edu). Another noteworthy influence during this period was work by the Office of Special Education Programs (OSEP), U.S. Department of Education. OSEP had established a specific grant program supporting research and development in technology, media, and materials for individuals with disabilities. With extensive input from consumers and professionals, OSEP facilitated the development of a research agenda to guide the federal grant process related to technology, media, and materials for students with disabilities (Schiller, 1993). Two subsequent reports describe the impact of the federal investment in special education technology research and development (Hauser & Malouf, 1996; U.S. Department of Education, 2000).

Considerable research on special education technology has been conducted during the past twenty years. Several literature reviews provide important historical snapshots concerning the effectiveness of specific technologies and interventions (Edyburn, 1995, 2000b, 2001a; Okolo, Bahr & Rieth, 1993; Woodward & Cuban, 2001, Woodward & Rieth, 1997). Generally, we've learned that the issues associated with capturing the potential of technology are much more complex than we original thought. Despite a knowledge base that demonstrates the effective use of technology by individual and small groups of students, we have not been able to systematically capture the benefits of technology on a large scale. As a result, significant numbers of individuals with disabilities who could benefit from technology have yet to be introduced to the possibilities. Therefore, it may be important to revisit the notion of a defining a technology research agenda (Pugach & Warger, 2001).

Early in the twenty-first century, a number of issues illustrate the need for concerted efforts to enlarge and systematize the research agenda concerning technology applications in special education. The following sections introduce an array of issues in six areas: accessibility, assistive technology, professional development, instructional technology, service delivery, and legal/policy issues, which currently impact the application of technology in special education. Within each topical area, key work summarizing the current knowledge base are identified. Emphasis will be placed on understanding the gaps in the current knowledge base in order to identify important questions that deserve additional research.

ACCESSIBILITY

While recognizing the power and potential of technology, the special education community has also discovered that many technology products, as they come out of the box, are inaccessible for individuals with disabilities. The response to these design shortcomings has generally been two-fold: (a) create modifications that work around the problem, and (b) educate designers on universal design strategies that address the needs of individuals with disabilities while improving functionality for everyone.

What Do Know About Accessibility?

Much of the early work in special education technology during the 1980s focused on adapting the computer to make it accessible for individuals with disabilities (Behrmann, 1988; Budhoff, Thormann & Gras, 1984; Burkhart, 1980; Lewis, 1993). Today, over 20,000 assistive technology products are available (AbleData, 2000) that allow an individual with a special need to operate a computer using alternatives to the standard interface (keyboard, mouse, screen, printer) using tools like voice input, alternative keyboards, switches, and screen readers. However, the process of identifying, evaluating, and selecting assistive technology is often time-consuming and dependent upon specialized expertise provided by technology specialists, augmentative communication specialists, and/or rehabilitation engineers.

As the Internet evolved, the extensive network of people involved in making computers accessible for individuals with disabilities, readily recognized that inadequate web design could disenfranchise the disability community. As a result, leaders were quick to call for design standards that would facilitate the use of assistive technologies like screen readers and enable users to personalize settings by over-riding default settings established by a designer (e.g. change the font size, alter the contrast between the text and the background). To a great extent, these efforts have been successful as a result of a multi-facted strategy to (a) impact web design standards (http://www.wai.org), (b) disseminate accessibility audit tools like Bobby (http://www.cast.org/bobby/), and (c) advance examples that illustrate good design is beneficial for everyone (Paciello, 2000; TRACE Center, 2001).

Some important breakthroughs in disability access computing occurred during the 1990s as a result of the research and development work of the TRACE Center at the University of Wisconsin-Madison. Recognizing the tremendous efforts required to retrofit assistive technology, the TRACE Center staff worked with computer manufacturers to include accessibility control panels within the system software on every computer. Hence, for selected types of impairments, accessibility is available on all computers by simply activating features in a control panel. This was a huge paradigm shift as it demonstrated that accommodations could be made proactively and without extensive specialized expertise. These initial efforts to improve access through better design contributed to what has come to be known as "universal design." Principles of universal design challenge designers to plan for the full continuum of diversity so that everyone may use a product without additional modifications or adaptations.

Recently, the concepts of universal design and access to the general education curriculum have converged into what is referred to as "universal design for learning" (Dolan, 2000; Orkwis & McLane, 1998; Rose, 2000; Rose, Sethuraman & Meo, 2000). A leader in the area of universal design for learning has been the Center for Applied Special Technology (CAST). In 1999, CAST received a five-year grant from the Office of Special Education Programs to establish the Center for Accessing the General Education Curriculum (http://www.cast.org/naec/). In their view, universal design is a critical issue if students with disabilities are going to be able to access the general education curriculum. As part of their work, CAST sponsors the National Consortium on Universal Design for Learning (http://www.cast.org/udl/). The fundamental challenge is to design educational environments and instructional materials in ways that will be accessible to all students entering the classroom: students unable to manipulate the pages of a textbook, students unable to read at grade level, students whose first language is not English, etc.

What Do We Need to Know About Accessibility?

The issue of accessibility is a central one for special education. Current advocacy efforts have increased knowledge of the issues involved in physical and sensory access and new awareness is emerging concerning cognitive access. Designs which explicitly plan for diversity and thereby reduce or eliminate the need for reactive accommodations are a promising development. However, much more needs to be done. A research agenda in the area of accessibility may explore any number of questions: What percentage of accessibility problems can be addressed through improved awareness of existing solutions (i.e. activating control panels, purchasing off-the-shelf products) versus problems that require individual evaluation, custom solutions, or extensive consultation? How can protocols be used by individuals and organizations to conduct proactive, self-audits of the accessibility of a learning environment (i.e. computer lab, distance education system, cdrom-based learning materials)? Given products created using principles of universal design for learning theory, what impact do they have on students? teachers? learning outcomes? How can technology be used to cognitively rescale information to make it accessible to individuals requiring more or less challenge?

ASSISTIVE TECHNOLOGY

Historically, the emphasis on technology for individuals with disabilities has been thought of as assistive technology, that is, extending the abilities of an individual in ways that provides physical access (i.e. wheelchairs, braces), and sensory access (i.e. Braille, closed captioning). Indeed, the legal definition of assistive technology is considerably broad:

> §300.5 Assistive technology device.
> As used in this part, Assistive technology device means any item, piece of equipment, or product system, whether acquired commercially off the shelf, modified, or customized, that is used to increase, maintain, or improve the functional capabilities of a child with a disability (Authority: 20 U.S.C. 1401(1)).

More recently, we've come to understand that additional attention must also be given to the use of technology for teaching and learning (Blackhurst & Lahm, 2000). In this chapter we'll use the term "special education technology" to cover both dimensions of assistive technology and instructional technology. Regardless of the specific application of technology, the general goal is always the same: to harness the potential of technology in ways that offer an individual with a disability increased opportunities for learning, productivity, and independence-opportunities that otherwise would not be available.

What Do We Know About Assistive Technology?

Compared to most areas of special education technology, we know much about assistive technology. In one of the major works within the discipline, Cook and Hussey (1995) examine the context of human performance, disability, and assistive technology and provide theoretical and practical insights into the ways technology enhances performance. Research syntheses completed by Thorkildsen (1994) offer a historical benchmark concerning the use of assistive technology, typically involving mobility and communication. Our understanding of assistive technology and specific disabilities has been increased through the work of Wehmeyer (1999) who examined mental retardation and Edyburn (2000a) who examined mild disabilities.

A number of studies contribute to the knowledge base concerning the acquisition and use of assistive technology. Todis and Walker (1993) and Todis (1996) helped us understand user perspectives on assistive technology. Research commissioned by United Cerebral Palsy (1999) offers insight into the technology needs of families and ways in which family centers can design programs and services to meet those needs. Parette and colleagues (Parette, 1999; Parette & Hourcade, 2000; Parette & Hourcade, 1997) have advanced a thoughtful series of inquiries into multicultural considerations and assistive technology. The state of assistive technology services in public schools has been examined by several research groups and contributes to an understanding of the variables impacting teacher support of students who use assistive technology (Derer, Polsgrove & Rieth, 1996; Hutinger, Johanson & Stoneburner, 1996; McGregor & Pachuski, 1996). Research by Raskind, Higgins, Slaff & Shaw (1998) has provided a glimpse of assistive technology use at home by students with learning disabilities and has the potential to offer important insights about the relationship of home-school computing much like the earlier work of Giacquinta, Bauer and Levin (1993).

An emerging topic of interest in the area of assistive technology focuses on issues involved in outcomes and quality. Smith (1996, 2000) has called attention to an array of measurement issues associated with understanding the impact of assistive technology. Zabala & Korsten (1999) have suggested that a series of changes can be expected when assistive technology is used effectively: quality, quantity, accuracy, rate, frequency, spontaneity, independence, and other. Their work offers an important starting point for creating a framework to document the effective use of assistive technology. Another promising development comes from the work of the QIAT Consortium (http://www.qiat.org/) which has sought to operationally define quality indicators for assistive technology services (QIAT Consortium, 2001). Finally, the results of a national survey of assistive technology outcome practices has been published in a three-volume series by RESNA (1998) that offers a snapshot of current instruments, resources, and strategies for assessing assistive technology outcomes.

What Do We Need to Know About Assistive Technology?

Significant gaps remain in what we know about assistive technology. For example, some questions that outline what we need to know, include: What is the current incidence of student use of assistive technology in schools? Are there key benchmarks to monitor concerning the equitable identification of need, selection, acquisition, and use of assistive technology? What impact does the use of assistive technology have on the academic performance of its users? Which current practices lead to effective and sustained use of assistive technology and which contribute to abandonment?

PROFESSIONAL DEVELOPMENT

Teachers play a critical role in supporting students using assistive and instructional technology (McGregor & Pachuski, 1996; U.S. Congress, Office of Technology Assessment 1988, 1995). As a result, considerable effort has been devoted to: (a) understanding what teachers need to know and be able to do; and (b) effective strategies for providing professional development to preservice and inservice teachers. Particularly troublesome is the fact that the need for extensive initial teacher preparation and on-going professional development in special education technology far exceeds our current ability to provide it.

What Do Know About Professional Development and Technology?

Lahm (Lahm, 1999, 2000) has reported on the efforts of the Technology and Media (TAM) Division of the Council for Exceptional Children to identify and validate the technology competencies expected of beginning special education teachers, master teachers, and assistive technology specialists. This work has been used extensively by universities and inservice providers as they design coherent training sequences. Alternatively, programs like RESNA (http://www.resna.org) and CSUN (http://www.csun.edu/cod/) have established national certification programs to certify participants, upon completion, as assistive technology specialists. These two programs fill an important gap in leadership development for the discipline as a recent review of personnel preparation programs in assistive technology by RESNA (1998) identified only 21 programs in the United States that provided coursework leading to a certificate or degree in special education technology. In practice, this means that individuals desiring training in assistive technology will find less than one program for every two states in the country. To say that the pipeline preparing new special education technology specialists is severely constricted may be an understatement.

The literature also documents efforts to explore innovative uses of technology to enhance technology training for special education teachers. Edyburn (2001b) outlines a taxonomy of technology-based training approaches and supports. Feit (1999) describes the important paradigm shift involved in the creation of the Intellitools Learning Activities Exchange. This online component of their web site is a place where teachers can download instructional materials and software players to use adapted curriculum materials with their students before devoting the time and effort to learning how to modify or create original materials of their own. By completely inverting the training sequence, Intellitools has discovered they should reprioritize their efforts and yet, accomplish far more and reach many more teachers. The Special Needs Opportunities Windows (SNOW) project at the University of Toronto has demonstrated the value of distance education as an effective means of providing inservice professional development on special education technology (http://snow.utoronto.ca/coursereg.html). The use of distance education to simultaneously help teachers learn about special education and gain important technology skills has also been explored by several groups of researchers (Blackhurst, Hales & Lahm, 1998; Smith, Smith & Boone, 2000).

The importance of self-directed learning for teachers and technology specialists cannot be overlooked. Professional organizations like TAM (http://www.tam.cec.org), RESNA (http://www.resna.org), and the newly formed Special Education Technology Special Interest Group (SETSIG) in the International Society of Technology in Education (ISTE) (http://www.iste.org) provide opportunities to join in membership with other professionals.

Resource organizations also play an important role in ongoing professional development given the variety of information products and services they provide (Closing the Gap, http://www.closingthegap.com; Center for Applied Special Technology, http://www.cast.org, Alliance for Technology Access, http://www.ataccess.org).

National special education technology conferences provide important opportunities for learning about new developments within the discipline. Major conferences include: ATIA (January, http://www.atia.org), TAM (January, http:/

/www.tam.cec.org), CSUN (March, http://www.csun.edu/cod/), NECC (June, http://www.neccsite.org), and Closing the Gap (October, http://www.closingthegap.com).

A final component of self-directed learning involves using key publications within the discipline to stay current. The following publications are among the core journals of the discipline: *Assistive Technology, Closing the Gap, Journal of Special Education Technology, Special Education Technology Practice*, and *Technology and Disability.*

What Do We Need to Know About Professional Development and Technology?

Despite the extensive knowledge base informing the design and delivery of professional development programs for teachers, a number of questions require additional study: What are the professional development habits of special education technology specialists? Which current professional development practices relative to technology and staff development can be documented as effective and which are known to have minimal impact? How much time should be devoted to maintaining current awareness and improving knowledge and skills in order for professionals to remain actively engaged in the knowledge base of the profession? What is the relationship between staff development concerning innovative applications of technology and student outcomes?

INSTRUCTIONAL TECHNOLOGY

Much of the original work on microcomputer use in special education centered on drill and practice learning activities. Moving beyond this remedial application of technology required a transformation in thinking and professional practice (Russell, Corwin, Mokros & Kapisovsky, 1989). Today, another transformation in thinking and professional practice is required as teachers and technology specialists must reconsider the role of technology in the inclusive classrooms, due to the fact that most students with disabilities spend a majority of their school day in general education classrooms. In response, the discipline of special education technology must place more emphasis on understanding and supporting the use of technology to enhance teaching and learning in additional to its traditional emphasis on assistive technology.

What Do We Know About Instructional Technology?

Several analyses have summarized what is known about the effective use of technology by students with disabilities (Okolo, Bahr & Rieth, 1993; Woodward & Rieth, 1997). Other insights concerning effective instructional technology practices may be gained from studies involving general education students (Fletcher-Finn & Gravatt, 1995; Roblyer, Castine & King, 1988; Schacter & Fagnano, 1999; Sivin-Kachala & Bialo, 1995, 2000). The evidence clearly suggests we know a lot about the use of technology to enhance writing and problem solving and have many examples of how technology can facilitate student achievement through its attributes of innovation, feedback, and engagement.

Two trends presently impacting classroom instructional practice are standards-based instruction (McLaughlin, Nolet, Rhim & Henderson, 1999) and high-stakes testing (Elliot, Erickson, Thurlow & Shriner, 2000). As a result, teachers perceive a clear need to ask questions related to the educational outcomes regarding the materials they select and the instructional approaches they use. Several recent works have important implications for using technology to help students with disabilities achieve high academic standards: curriculum design strategies (Burke, Hagan & Grossen, 1998), grading modifications (Christensen & Vogel, 1998; Welch, 2000), cognitive credit cards (Edmunds, 1999), and validated instructional strategies (Sikorski, Niemiec & Walberg, 1996). New accountability tools that help schools align standards, learning experiences, and technology in ways that document students' achievement of high academic standards are an important development (U.S. Department of Education, 1998).

The issue of technology integration is fundamental to understanding the use of technology to enhance instruction. The difficulties associated with technology integration are well documented: lack of teacher time; access to hardware, software and support; limited leadership, lack of of a common vision or rationale for technology use; limited training and support; and the impact of current assessment practices on defining what teachers must teach and that what students learn with technology may not be readily measured on standardized tests (U.S. Congress, Office of Technology Assessment, 1995).

Willis (1993) adds a number of other interesting dimensions of the problems teachers will confront when they try to integrate technology: curricular integration is a complex, difficult-to-learn process; many educators feel isolated and alone; time to experiment, explore, and study innovations is essential but rare in schools; top down projects tend to fail over time; resentment and resistance destroys projects, ownership is critical to success;

bottom-up projects tend to fail over time; administrative support is critical; nonexistent, inadequate, or inconsistent support is a major reason for failure; and theories of change are useful planning guides for change. Finally, experienced technology using teachers conclude – at least initially – most uses of computers make teaching more challenging and require more effort (U.S. Congress, Office of Technology Assessment, 1988, 1995). Individually, and collectively, the impact of these factors suggest the true complexity of the challenge associated with integrating technology into the curriculum.

What Do We Need to Know About Technology Integration?

Despite significant work in the area of technology integration, a number of factors impact our ability to make integration happen on a systemic basis. Therefore, a number of questions are worthy of additional research: How does technology enhance teaching? How does technology enhance learning? When students use technology to enhance their productivity as a learner, what types of outcomes do they report? What role can technology play in facilitating the work of teachers to make modifications in curriculum, instruction, and/or assessment for students in general education settings? What types of measurement and reporting systems should document claims of the effectiveness of technology-based instructional interventions?

SERVICE DELIVERY

Issues involved in the delivery of assistive technology services in schools are likely to gain additional attention in the future. At the present time, there is little evidence to suggest that assistive technology consideration is happening as required by law. As a result, in the near future we are likely to see new emphasis on service delivery systems equivalent to child find (screening for assistive technology needs) and prereferral interventions (equipping classrooms with common devices that have been demonstrated to help students thereby reducing the need for extensive assistive technology evaluations).

What Do We Know About Special Education Technology Service Delivery Systems?

In a new review of the literature on special education technology service delivery systems, Ludlow (2001) provides a comprehensive analysis of promises and pitfalls associated with technology and special education teacher education. This review fills a significant void on this topic.

In work that has important implications for both research and practice, Haines and Sanche (2000) reviewed four common assistive technology assessment models. Subsequently, they describe a normalization process used to standardize the models and terms, and sequence the components advanced by each theorist. The result is a synthesis of the four individual models into a coherent framework, "The AT CoPlanner Model," which they have implemented as a content module for the software product, "CoPlanner."

To-date, few alternatives (Anson, 1997) have emerged to a service delivery system that is labor intensive and requires specialized expertise not commonly found in every community. As a result, it can be surmised that the size of the underserved population is quite significant.

The lack of empirical data on special education technology service delivery methods requires the use of anecdotal observations to describe current practices:

- It is common practice to utilize a multidisciplinary team made up of teachers, technology specialists, occupational therapist, speech therapist, and physical therapist when conducting an assistive technology evaluation.
- Most individuals participating in an assistive technology team do so as a result of part-time release from their regular case/teaching load.
- Most assistive technology evaluations involve an in-depth evaluation that, in many respects, mirrors the special education referral process, and therefore, is perhaps comparable in terms of cost, time involved, and efficiency.
- There is little evidence to suggest that schools use any systemic screening process to identify students who potentially could benefit from assistive technology.
- Students who have access to assistive technology often do so as the result of advocacy efforts that challenge the system rather than through a systemic process that ensures that all students in need of devices have them.
- Of the many students who currently use assistive technology, the most common applications involve technologies that overcome physical challenges or enhance communication abilities.
- Most AT teams lament that the majority of their time is spent assessing new students for assistive technology needs rather than being engaged in on-going support and follow-up of current assistive technology-using students. Implicitly, the emphasis is on acquisition (i.e. shopping), rather than on implementation and assessment of outcomes.
- Per federal mandate, every IEP team is required to document their efforts to consider assistive technology. As it pertains to students with mild disabilities (LD, ED, MR), ages 6–21, this effects over 3.8 million students. Current

assistive technology service delivery systems, developed to respond to the needs of students with low incidence disabilities, do not appear capable of being scaled-up to meet the needs of students with high incidence disabilities.

What Do We Need to Know About Special Education Technology Service Delivery Systems?

Little research has been conducted on the service delivery systems used by schools and state education agencies to provide special education technology. As a result, there is much to be learned. Some questions that should be addressed in future research studies, include: What organizational structures are used to organize assistive technology services in schools? How many certified assistive technology specialists are employed in public and private schools? full-time? part-time? What is the composition of assistive technology teams in public schools? How long does it take to assess the need for assistive technology, acquire devices, train, and implement?

LEGAL/POLICY ISSUES

As noted at the beginning of this chapter, the field of special education technology has been profoundly influenced by the legislative action that sought to realize the potential of technology for persons with disabilities. As a result, it has become increasingly important that policy studies become part of the research agenda for the discipline.

What Do We Know About Legal/Policy Issues?

The issue of assistive technology consideration is a rather recent development. Its origin can be traced to the Individuals with Disabilities Education Act Amendments of 1997 (Public Law 105-17) which contained a requirement for the Individual Education Program (I.E.P.) teams to consider assistive technology in the development of an IEP: "The IEP Team shall – (v) consider whether the child requires assistive technology devices and services." [Section 614 (d)(3)(B) Consideration of Special Factors.]

Whereas some observers believe this language reflects a new federal policy, Golden (1998) argues that it simply formalizes a previous responsibility:

> The IDEA requires schools to provide AT if it is needed for a student to receive a free appropriate publication education (FAPE). FAPE can include a variety of services such as special education, related services, supplementary aids and services, program modifications or support for school personnel. AT, just like other components of FAPE, must be provided

> at no cost to parents. The specific IDEA requirement for schools to provide AT is as follows:
>
> § 300.308 Assistive Technology
> Each public agency shall ensure that assistive technology devices or assistive technology services or both, as those terms are defined in 300.5 – 300.6 are made available to a child with a disability if required as part of a child's (a) Special education under 300.17; (b) Related services under 300.16; or (c) Supplementary aids and services under 300.550(b)(2)" (p. 4).

Golden's analysis highlights a critical issue: Free Appropriate Public Education (FAPE). Schools are required to provide assistive technology for students that need such tools, if they are necessary, for the student's participation in and benefit from a free appropriate public education (Etscheidt & Bartlett, 1999; Huefner, 2000). The historical implications of this requirement are unquestioned in the context of mobility (i.e. a powered wheelchair) and communication (i.e. an augmentative communication system). However, the requirement covers all disabilities and therefore issues like the following have emerged: "Jimmy's handwriting is not legible, therefore he needs a laptop computer." While such a claim and solution may indeed be certified by an I.E.P. team, the budgetary implications of this mandate, when applied to a high incidence population, has created an environment where administrators are reluctant to approve requests for assistive technology for students with mild disabilities given the fact that they have 50 students like Jimmy within their building (Edyburn, 2000a). Of course, interventions other than a laptop computer may also be appropriate.

Has technology been considered? A guide for IEP teams (Chambers, 1997) is an acknowledged key resource on the topic of assistive technology consideration. This book is an outcome of the author's research which involved a delphi study of assistive technology experts and focus groups with trainers and consumers of assistive technology services in response to her observation that the 1997 reauthorization of IDEA required that assistive technology be considered, but the legislation offered no guidelines on how to implement the requirement. A valuable component of Chamber's work is a flow chart of questions that should be asked and answered, by an I.E.P team. As a result of engaging in the process, she argues that teams will automatically generate the documentation of their assistive technology consideration efforts on behalf of a child.

What Do We Need to Know About Legal/policy Issues?

As school districts seek to implement assistive technology on a systemic basic, a recent analysis by Golden (1999) provides benchmark estimates of the percentage of students by disability category that could potential benefit from assistive technology. These expectance figures should be confirmed or modified through research as an important policy tool that spurs action for reaching the underserved.

New research on service delivery models needs to examine alternatives to current models and examine universal access for all who could benefit. Obviously, funding is still a problem. Additional questions worthy of exploration include: Do all students who could potentially benefit from assistive technology have access to appropriate devices and services? Given a shortage of specialized personnel and the size of the high incidence disability population, what types of assistive technology interventions can be shown to be effective for students with mild disabilities? How might toolkits with these interventions be distributed and used as a pre-referral intervention? What factors should be considered when evaluating a due process claim concerning the need for assistive technology? What types of decision aids can be shown to improve decision making relative to the consideration, selection, and use of assistive technology?

CONCLUDING COMMENTS

Despite the impact of two powerful forces throughout the twentieth century, innovation and policy, and the success in demonstrating that a technological device or intervention works for an individual or a small group, the field of special education technology has struggled with the "scaling-up challenge." That is, how do we reach all the individuals that could potentially benefit from using technology? The gap between the potential of technology and current practice has been a source of frustration to many parents, professionals, and policy makers. Ideally, research should advance solutions to the difficulties associated with the "scaling up challenge."

As future leadership and research agendas are developed, consideration should be given to three aspects of using research to enhance decision-making concerning special education technology. Clearly, more research on the effective use of technology is urgently needed. Indeed, many more products are currently available in the marketplace, and are finding their ways into schools, than products with a record of research-validated evidence of effectiveness. Second, significant attention must be devoted to ensuring that research results are relevant. Many large multi-year projects currently take two to three years to collect data, two additional years to have the results published, and end up describing the effectiveness of products that are no longer available or have been updated through several revision cycles. Finally, increased emphasis on

Table 1. Critical Unanswered Questions in Special Education Technology Research.

Accessibility

- What percentage of accessibility problems can be addressed through improved awareness of existing solutions (i.e. activating control panels, purchasing off-the-shelf products) versus problems that require individual evaluation, custom solutions, or extensive consultation?
- How can protocols be used by individuals and organizations to conduct proactive, self-audits of the accessibility of a learning environment (i.e. computer lab, distance education system, cdrom-based learning materials)?
- Given products created using principles of universal design for learning theory, what impact do they have on students? teachers? learning outcomes?
- How can technology be used to cognitively rescale information to make it accessible to individuals requiring more or less challenge?

Assistive Technology

- What is the current incidence of student use of assistive technology in schools?
- Are there key benchmarks to monitor concerning the equitable identification of need, selection, acquisition, and use of assistive technology?
- What impact does the use of assistive technology have on the academic performance of its users?
- Which current practices lead to effective and sustained use of assistive technology and which contribute to abandonment?

Professional Development

- What are the professional development habits of special education technology specialists?
- Which current professional development practices relative to technology and staff development can be documented as effective and which are known to have minimal impact?
- How much time should be devoted to maintaining current awareness and improving knowledge and skills in order for professionals to remain actively engaged in the knowledge base of the profession?
- What is the relationship between staff development concerning innovative applications of technology and student outcomes?

Instructional Technology

- How does technology enhance teaching? How does technology enhance learning?
- When students use technology to enhance their productivity as a learner, what types of outcomes do they report?
- What role can technology play in facilitating the work of teachers to make modifications in curriculum, instruction, and/or assessment for students in general education settings?
- What types of measurement and reporting systems should document claims of the effectiveness of technology-based instructional interventions?

Service Delivery

- What organizational structures are used to organize assistive technology services in schools?
- How many certified assistive technology specialists are employed in public and private schools? full-time? part-time?
- What is the composition of assistive technology teams in public schools?
- How long does it take to assess the need for assistive technology, acquire devices, train, and implement?

Table 1. Continued.

Legal/Policy Issues

- Do all students who could potentially benefit from assistive technology have access to appropriate devices and services?
- Given a shortage of specialized personnel and the size of the high incidence disability population, what types of assistive technology interventions can be shown to be effective for students with mild disabilities? How might toolkits with these interventions be distributed and used as a pre-referral intervention?
- What factors should be considered when evaluating a due process claim concerning the need for assistive technology?
- What types of decision aids can be shown to improve decision making relative to the consideration, selection, and use of assistive technology?

action research models and procedures may ultimately provide schools with the necessary data-based decision-making tools. Given the pace of change in the marketplace and the rate of adoption, action research can serve to assess the efficacy of innovative applications of special education technology in a timely and relevant manner.

The purpose of this chapter has been to provide a context for understanding the persistent effort of the profession to identify, implement, and evaluate applications of technology that serve to amplify and enhance communication, mobility, independence, and learning by individuals with disabilities. Critical issues associated with six areas of special education technology: accessibility, assistive technology, instructional technology, professional development, service delivery, and legal/policy issues, have been presented. Emphasis has been placed on summarizing key works and understanding the gaps in the current knowledge base in order to outline important questions that deserve further research. A list of the questions we need to know which have been advanced by the author are presented in Table 1. Readers are encouraged to consider the issues raised in this chapter as they define future leadership and research agendas and how their work may contribute important research findings to understanding critical questions about the effective use of technology in special education.

REFERENCES

AbleData. (2000). Search for assistive technology products. Available from: http://www.abledata.com/text2/search.htm

Anson, D. (1997). *Alternative computer access: A guide to selection.* Philadelphia, PA: F. A. Davis.

Behrmann, M. M. (Ed.) (1988). *Integrating computers into the curriculum: A handbook for special educators*. Boston: College-Hill.

Blackhurst, A. E. (1997). Perspectives on technology in special education. *Teaching Exceptional Children, 29*(5), 41–48

Blackhurst, A. E., & Edyburn, D. L. (2000). A brief history of special education technology. *Special Education Technology Practice, 2*(1), 21–36.

Blackhurst, A. E., Hales, R. M., & Lahm, E. A. (1998). Using an education server software system to deliver special education coursework via the World Wide Web. *Journal of Special Education Technology, 13*(4), 78–98

Blackhurst, A. E., & Lahm, E. A. (2000). Technology and exceptionality foundations. In: J. D.Lindsey, (Ed.), *Technology and Exceptional Individuals* (3rd ed., pp. 3–45). Austin, TX: ProEd.

Budhoff, M., Thormann, J., & Gras, A. (1984). *Microcomputers in special education*. Cambridge, MA: Brookline.

Burke, M. D., Hagan, S. L., & Grossen, B. (1998). What curricular designs and strategies accommodate diverse learners? *Teaching Exceptional Children, 31*(1), 34–38.

Burkhart, L. (1980). *Homemade battery powered toys and educational devices for severely handicapped children* (3rd ed.). College Park, MD: Author.

Chambers, A. C. (1997). *Has technology been considered? A guide for IEP teams*. Reston, VA: The CASE/TAM Assistive Technology Policy and Practice Group.

Christensen, J., & Vogel, J. R. (1998). A decision model for grading students with disabilities. *Teaching Exceptional Children, 31*(2), 30–35.

Cook, A.. M., & Hussey, S. M. (1995). *Assistive technologies: Principles and practice*. St. Louis, MO: Mosby-Year Book, Inc.

Derer, K., Polsgrove, L., & Rieth, H. (1996). A survey of assistive technology applications in schools and recommendations for practice. *Journal of Special Education Technology, 13*(2), 62–80.

Dolan, B. (2000). Universal design for learning. *Journal of Special Education Technology, 15*(4), 44–51.

Edyburn, D. L. (2001a). JSET 2000 in review: A synthesis of the special education technology literature. *Journal of Special Education Technology, 16*(2), 5–25.

Edyburn, D. L. (2001b). Training as a technology integration strategy. *Closing the Gap, 19*(6), 12–13.

Edyburn, D. L. (2000a). Assistive technology and students with mild disabilities. *Focus on Exceptional Children, 32*(9), 1–24.

Edyburn, D. L. (2000b). JSET 1999 in review: A synthesis of the special education technology literature. *Journal of Special Education Technology, 15*(1), 7–18.

Edyburn, D. L. (1995). Technology, learning disabilities, and ways of knowing: A review of recent doctoral dissertations. *LD Forum, 19*(1), 54–62.

Edmunds, A. L. (1999). Cognitive credit cards: Acquiring learning strategies. *Teaching Exceptional Children, 31*(4), 68–73.

Elliot, J. L., Erickson, R. N., Thurlow, M. L., & Shriner, J. G. (2000) State-level accountability for the performance of students with disabilities. *Journal of Special Education, 34*(1), 39–47.

Etscheidt, S. K., & Bartlett, L. (1999). The IDEA Amendments: A four-step approach for determining supplementary aids and services. *Exceptional Children, 65*, 163–174.

Fein, J. (1996). A history of legislative support for assistive technology. *Journal of Special Education Technology, 13*(1), 1–3.

Feit, S. (1999). Intellitools activity exchange: An interview with Suzanne Feit. *Special Education Technology Practice*, *1*(3), 20–21.

Fletcher-Finn, C. M., & Gravatt, B. (1995). The efficacy of computer assisted instruction (CAI): A meta-analysis. *Journal of Educational Computing Research*, *12*(3), 219–242.

Giacquinta, J. B., Bauer, J. A., & Levin, J. E. (1993). *Beyond technology's promise: An examination of children's educational computing at home*. NY: University of Cambridge Press.

Golden (1999). Assistive technology policy and practice. What is the right thing to do. What is the reasonable thing to do? What is required and must be done? *Special Education Technology Practice*, *1*(1), 12–14.

Haines. L., & Sanche, B. (2000). Assessment models and software support for assistive technology teams. *Diagnostique*, *25*, 291–305.

Hannaford, A. E. (1993). Computers and exceptional individuals. In: J. D. Lindsey (Ed.), *Computers and Exceptional Individuals* (2nd ed., pp. 3–26). Austin, TX: Pro-Ed.

Hauser, J., & Malouf, D. (1996). A federal perspective on special education technology. *Journal of Learning Disabilities*, *29*, 504–511.

Hobson, D. A. (1996). RESNA: Yesterday, today, and tomorrow. *Assistive Technology*, *8*(1), 131–143.

Huefner, D. S. (2000). The risks and opportunities of the IEP requirements under IDEA '97. *Journal of Special Education*, *34*(4), 195–204.

Lahm, E. A. (2000). Special education technology: Defining the specialist. *Special Education Technology Practice*, *2*(3), 22–27.

Lahm, E. A., & Nickels, B. L. (1999). What do you know? Assistive technology competencies for special educators. *Teaching Exceptional Children*, *32*(1), 56–63.

Lewis, R. B. (1993). *Special education technology: Classroom applications*. Pacific Grove, CA: Brooks/Cole.

Ludlow, B. L. (2001). Technology and teacher education in special education: Disaster or deliverance? *Teacher Education and Special Education*, *24*, 143–163.

McGregor, G., & Pachuski, P. (1996). Assistive technology in schools: Are teachers, ready, able, and supported? *Journal of Special Education Technology*, *13*(1), 4–15.

McLaughlin, M. J., Nolet, V., Rhim, L. M., & Henderson, K. (1999). Integrating standards: Including all students. *Teaching Exceptional Children*, *31*(3), 66–71.

Okolo, C. M., Bahr, C. M., & Rieth, H. J. (1993). A retrospective review of computer-based instruction. *Journal of Special Education Technology*, *12*(1), 1–27.

Orkis, R., & McLane, K. (1998). *A curriculum every student can use: Design principles for student access. OSEP Topical Brief*. Reston, VA: Council for Exceptional Children. Also available online at: http://www.cec.sped.org/osep/udsign.htm

Paciello, M. G. (2000). *Web accessibility for people with disabilities*. Lawrence, KS: RD Books.

Parette, P. (1999). Transition and assistive technology planning with families across cultures. *Career Development for Exceptional Individuals*, *22*, 213–231.

Parette, H. P., & Hourcade, J. J. (1997). Family issues and assistive technology needs: A sample of state practices. *Journal of Special Education Technology*, *13*(3), 27–43

Parette, H. P., & Hourcade, J. J. (2000). Assistive technology training for parents of students with disabilities. *Special Education Technology Practice*, *2*(2), 20–24.

Pugach, M. C., & Warger, C. L. (2001). How does technology support a special education agenda? Using what we have learned to inform the future. In: J. Woodward & L. Cuban (Eds), *Technology, Curriculum and Professional Development: Adapting Schools to Meet the Needs of Students With Disabilities* (pp. 226–239). Thousand Oaks, CA: Corwin Press.

QIAT Consortium (2001). Quality indicators for assistive technology services. *Special Education Technology Practice*, *3*(1), 14–15.

Raskind, M. H., Higgins, E. L., Slaff, N. B., & Shaw, T. K. (1998). Assistive technology in the homes of children with learning disabilities: An exploratory study. *Learning Disabilities: A Multidisciplinary Journal*, *9*(2), 47–56.

RESNA. (1999). Survey of assistive technology training programs. Available online from: http://www.resna.org/

Rehabilitation Engineering and Assistive Technology of North American (RESNA) (1998). Volume 1: RESNA Resource Guide for Assistive Technology Outcomes: Measurement Tools. Arlington, VA: Author.

Rehabilitation Engineering and Assistive Technology of North American (RESNA) (1998). Volume 2: RESNA Resource Guide for Assistive Technology Outcomes: Assessment Instruments, Tools, & Checklists. Arlington, VA: Author.

Rehabilitation Engineering and Assistive Technology of North American (RESNA) (1998). Volume 3: RESNA Resource Guide for Assistive Technology Outcomes: Developing Domains of Need and Criteria of Services. Arlington, VA: Author.

Roblyer, M. D., Castine, W. H., & King, F. J. (1988). Assessing the impact of computer-based instruction. *Computers in the Schools*, *5*(3/4), 1–149.

Rose, D. (2000). Universal design for learning. *Journal of Special Education Technology*, *15*(1), 67–70.

Rose, D., Sethuraman, S., & Meo, G. J. (2000). Universal design for learning. *Journal of Special Education Technology*, *15*(2), 56–60.

Russell, S. J., Corwin, R., Mokros, J. R., & Kapisovsky, P. M. (1989). *Beyond drill and practice: Expanding the computer mainstream*. Reston, VA: Council for Exceptional Children.

Schacter, J., & Fagnano, C. (1999). Does computer technology improve student learning and achievement? How, when, and under what conditions? *Journal of Educational Computing Research*, *20*, 329–343.

Schiller, E. (1993). The national agenda for the technology, media, and materials program for individuals with disabilities. *The TAM Newsletter*, *8*(1), 7–18.

Sikorski, M., Niemiec, R. P., & Walberg, H. J. (1996). A classroom checkup: Best teaching practices in special education. *Teaching Exceptional Children*, *29*(1), 27–29.

Sivin-Kachala, J., & Bialo, E. R. (2000). *2000 research report on the effectiveness of technology in schools* (7th ed). Washington, D.C.: Software Information Industry Association.

Sivin-Kachala, J., & Bialo, E. R. (1995). *Report on the effectiveness of technology in schools*, 1990–1994. Washington, D.C.: Software Publishers Association.

Smith, R. O. (2000). Measuring assistive technology outcomes in education. *Diagnostique*, *25*, 273–290.

Smith, R. O. (1996). Measuring assistive technology interventions: challenges and innovation. *Assistive Technology*, *8*(2), 71–81.

Todis, B. (1996). Tools for the task? Perspectives on assistive technology in educational settings. *Journal of Special Education Technology*, *13*(2), 49–61.

Todis, B., & Walker, H. M. (1993). User perspectives on assistive technology in educational settings. *Focus on Exceptional Children*, *26*(3), 1–16.

TRACE Center. (2001). Designing a more usable world – for all. Document available at http://www.trace.wisc.edu/world/

United Cerebral Palsy. (1999). Needs assessment and resource analysis for the family center on technology and disability. Washington, D.C.: author. Available at: http://fctd.ucp.org/fctd/tneedsassess.htm

U.S. Department of Education. (1998). *An educator's guide to evaluating the use of technology in schools and classrooms*. Available at: http://www.ed.gov/pubs/EdTechGuide/

U.S. Congress, Office of Technology Assessment (1995). *Teachers and technology: Making the connection*. Washington, D.C.: U.S. Government Printing Office. Also available at http://www.wws.princeton.edu:80/~ota/disk1/1995/9541_n.html

U.S. Congress, Office of Technology Assessment (1988). *Power on: New tools for teaching and learning*. Washington, D.C.: U.S. Government Printing Office. Also available at: http://www.wws.princeton.edu:80/~ota/ns20/catte_n.html

U.S. Congress, Office of Technology Assessment, (1982). *Technology and handicapped people*. Washington, D.C.: Author.

U.S. Department of Education. (2000). *Twenty-first annual report to Congress on the implementation of the Individuals with Disabilities Education Act*. Washington, D.C.: Author. Available online: http://www.ed.gov/offices/OSERS/OSEP/OSEP99AnlRpt/

Wehmeyer, M. L. (1999). Assistive technology and students with mental retardation: Utilization and barriers. *Journal of Special Education Technology, 14*(1), 48–58.

Welch, A. B. (2000). Responding to student concerns about fairness. *Teaching Exceptional Children, 33*(2), 36–40.

Willis, J. (1993). What conditions encourage technology use? It depends on the context. *Computers in the Schools, 9*(4), 13–32.

Woodward, J., & Cuban, L. (2001). *Technology, curriculum and professional development: Adapting schools to meet the needs of students with disabilities*. Thousand Oaks, CA: Corwin Press.

Woodward J., & Rieth, H. (1997). A historical review of technology research in special education. *Review of Educational Research, 67*(4), 503–536.

Zabala, J. S., & Korsten, J. E. (1999). Beyond "try it! you'll like it . . . or maybe you won't?": Making a measurable difference with assistive technology. Preconference workshop handout. 1999 Closing the Gap Conference. Minneapolis, MN.

AN EXPLORATION OF SYSTEMS THINKING, TEACHER CHOICE, AND REQUIREMENTS FOR IMPLEMENTING TECHNOLOGY INTO CLASSROOMS FOR STUDENTS WITH EMOTIONAL AND LEARNING DISABILITIES

John Castellani and Michael Behrmann

ABSTRACT

This chapter examines the role of the Internet in a system of self-contained schools for students with emotional and learning disabilities and the adoption of the Internet as a tool for teaching and learning. This qualitative study is part of a larger research project that explored issues for teaching and learning with the Internet for students with emotional and learning disabilities. The following research identifies how school context influenced teachers' use of technology tools and the Internet through teacher and administrator perceptions of the change process. Results from this study suggest that: (a) assistive and instructional technologies are necessary tools used to accommodate students with emotional and learning disabilities; (b) documentation on advances in classroom instruction around emerging technologies needs to reflect appropriate practice and involve all stakeholders, (c) students with

Technological Applications, Volume 15, pages 119–148.
2001 by Elsevier Science Ltd.
ISBN: 0-7623-0815-x

emotional and learning disabilities are increasingly at-risk for becoming part of the gap of those who do not have access to technology, and (d) systems implementing technology innovations for students with learning and emotional disabilities need to consider decision-making processes and responsibilities for systemic change in therapeutic technology-based interventions.

Special education systems are expanding as advances in legal requirements, changing technologies, and alternative teaching ideas influence classroom settings (Behrmann, 1998). The authors of IDEA '97 acknowledged the role that instructional and assistive technology (IT and AT) has for assisting students with disabilities to meet their educational goals. As a result, schools are fulfilling their legal obligation to provide appropriate technology to support teaching and learning (Julnes & Brown, 1993). Educators and rehabilitation professionals have historically used assistive technology to accommodate individuals with severe physical disabilities, speech disorders, and sensory impairments. The reauthorization of IDEA (1997) legislation has extended the term assistive technology to include a broader range of instructional technologies. By expanding the definition of assistive technology, educators have greater opportunities for exploring a range of technology accommodation options for students with high incidence disabilities. Legal requirements for accommodating students with technology now include instructional technologies necessary to individualize learning environments for students with mild mental retardation, and specific learning and emotional disabilities.

IDEA '97 legislation specifically requires that teams developing Individual Education Programs (IEPs) purposefully consider a student's IT and AT needs across disabilities to include any device or service necessary for educational access and progress. For students with learning and emotional disabilities, the presence of digital electronic materials and the influences of information technology on teaching and learning have emerged. There is a current movement within education to provide textual information in universally accessible formats necessary for individualization and accommodation (Orkwis & Mclane, 1998). In many cases, traditional text-based materials are now available within CD-ROMs and on the Internet. Technology support tools are being used to assist students with learning and emotional disabilities access and use these digital materials (Castellani & Jeffs, 2001). As schools adopt these IT options, there are increasing challenges for considering broader technology-based systems issues for supporting students and teachers in the classroom.

This chapter examines the role of the Internet in a system of self-contained schools for students with emotional and learning disabilities. This chapter explores the adoption of the Internet as a tool for teaching and learning and discusses the system-wide issues that resulted from this innovation. This research is part of a larger study that explored issues for teaching and learning with the Internet and the role of professional development for supporting teachers learning about new technology tools. The following research identifies how school context influenced teachers' use of technology tools and the Internet. Results of this study are presented as well as a discussion of the implications, to include: (1) teacher and student access to technology and curriculum resources; (2) systemic adoption of innovations and teacher choice; and (3) appropriate documentation and transference of new knowledge.

IMPLICATIONS FOR AN EMERGING INFORMATION-BASED SOCIETY

As society changes, there is a pressing need to accommodate all students with disabilities with the IT and AT necessary for competing in our current "highly technical and networked" global economy. This need is continuously reinforced by the marked increase in networking and telecommunications technologies available for classroom instruction. According to the National Center for Educational Statistics (1999), 95% of all U.S. schools and 63% of instructional classrooms have access to the Internet, and this number is growing. The NCES reports that large schools are more likely to have access to the Internet than small schools. Public schools in high poverty areas are less likely to be connected to the Internet than schools in affluent areas. Among all public schools, the NCES documents that 20% of teachers used advanced telecommunications for teaching; however, only 13% of the schools reported that training was available for use of school networks. For regular education teacher preparation and in-service training, there has been a unilateral response to train teachers to work with technology and telecommunications in the schools and prepare students for a networked society. As IT grows, many educators are paying close attention to the digital divide and putting systems in place to deal with those who do not have access to technology. The U.S. Department of Commerce (2000) reports that access to the Internet has increased to 50% in homes across the country. However, "persons with a disability are only half as likely to have access to the Internet as those without a disability: 21.6% compared to 42.1%. And while just under 25% of those without a disability have never used a personal computer, close to 60% of those with a disability fall into that category" (U.S. Department of Commerce, 2000).

Technology Access and Curriculum Resources

Increasing access to digital education materials has encouraged new thinking about universal design for learning. Accessibility to digital general education materials has become an important part of curriculum modification. Teachers, software developers, and others have initiated the development of multimedia supported teaching and learning environments that can be controlled by the user and utilized when students need multi-modal input for identified authentic and/or discreet learning tasks (CAST, 2000).

Understanding new pedagogical issues for students with learning and emotional disabilities is essential to advances in curriculum delivery. New ideas about activities oriented teaching, anchored instruction, and constructivist-learning environments have encouraged the development and use of teaching strategies outside of traditional text-based information delivery systems (Mastropieri & Scruggs, 1994; Patton, 1995; Cognition and Technology Group, 1994; Brandt & Perkins, 2000). New technologies have proven to accommodate learning styles and promote individualized instruction. These technologies include speech synthesizers, voice recognition, word prediction, hypermedia, hypertext, and virtual worlds (Dede, 1998; Hasselbring, 1994; Hasselbring, 1989; Inman, 1997; Jacobson & Spiro, 1995; MacArthur, 1997). Additionally, Instructional (IT) and assistive technologies (AT) have positively influenced the learning outcomes for persons with severe physical disabilities, cognitive and sensory impairments, and learning disabilities, from preschool to post-secondary schooling (e.g. Behrmann, 1998; Higgins & Boone, 1997; Holzberg, 1995; Lewis, 1993; Male, 1997; Raskind, 1999; Vanderheiden, 1985). IT has specifically influenced teaching and learning for students with high incidence disabilities by serving as an effective motivator, providing opportunities for cooperative learning opportunities, and enhancing students' ability to engage in self-monitoring activities (Castellani, 2000; Fitzgerald, 1990; Fitzgerald & Coury, 1996). In these cases, technology has influenced learning by increasing student self-esteem and facilitating self-expression. Research on the use of IT and AT for teaching and learning continues to influence new understanding for teaching and learning; however, external influences on technology-based innovations needs our continuous focus.

SYSTEMIC ADOPTIONS OF TECHNOLOGY-BASED INNOVATIONS

The literature has documented the effect that educational systems have on technology use in the classroom, including time for teacher planning, administrative support, links to university environments, ongoing staff development and training, and the integration of innovative teaching and learning strategies (ISTE, 1997; OSEP, 1997; U.S. Department of Education, 1996). Dede (1999) states, "Without undercutting their power, change strategies effective when pioneered by leaders in educational innovation must be modified to be implemented by typical educators." More holistic views and interactions between teachers and school administrators are needed for technology to impact school settings, accessible technology-based curriculum development and instructional strategies, and ultimately student learning (Pugach, 1996; Rosenberg, Griffin, Kilgore & Carpenter, 1997; CAST, 2000). Much of this depends on how school systems interpret new tools and allocate time, energy, personnel, and training to support instructional programs (OTA, 1995; Norton, 1997).

Many reform efforts in special education have focused primarily on equal educational access to technology, including the Internet, for individuals with disabilities (OSEP, 1997). Within the context of technology and growth within the past twenty years, the field of special education is one that has been defined by additional public and private reforms. Because of federal legislation and entitlement programs, increasing numbers of students with disabilities attend public and private schools supported by public funds. The conflict between the prescriptive nature of special education law and constitutional law within the states to determine educational requirements for students with disabilities has resulted in a continual shift in the patterns of delivery systems and specific placement criteria. As a result, programs for teaching and learning are often affected by how individuals at the state and local level, directs, funds, and determines the education of students with disabilities. Laws continue to expand the specifications and requirements to provide appropriate individualized teaching strategies to meet the learning needs of students with disabilities.

Larry Cuban (1979) argues that the integration of new ideas into the curriculum has more to do with the level of teacher autonomy than it does teacher's willingness or unwillingness to implement an innovation. He describes a teacher's ability to make instructional change as "situationally constrained choice." Identification of the context in which choices are made furthers an understanding of adapting new approaches for teaching and learning.

Teacher Choice and Technology

Larry Cuban (1993) explains that the regular and/or alternative education teacher is typically removed from many instructional decision-making processes necessary for implementing new classroom practices. Teachers rarely have the opportunity to decide the make-up of students in a given class, the content they are to teach, and how student behavior management will be handled school wide. Cuban describes these phenomena as situationally constrained choice.

Situationally constrained choice influences teacher autonomy by affecting the number of instructional changes that an individual can make about classroom practices as they are attempting to combine their own ideas about teaching and learning into larger systemic efforts. One way to discuss innovation is to think about what questions need to be answered before conclusions can be made about why teachers are or are not implementing new ideas and tools into their curriculum (Cuban, 1979). Administrative personnel typically address student groupings, student-teacher ratios, the curriculum followed for instructional purposes, administration of standardized tests, as well as the length of the school day, arrangements of periods, and the schedule of courses for teachers. As technology becomes integrated into school settings, networking issues, policies for appropriate student use of information technologies, and hardware placement become critical. Decisions about technology are typically made *outside* of classrooms. Autonomous teacher decisions are more specific to issues *inside* of the classroom, such as classroom arrangement, type of student activities, and the use of instructional tools. As the tools themselves become systems issues, questioning how systems address technology innovations is critical (Cuban, 1993; Fullan, 1991; Dede, 1998; Dede, 2000).

It is necessary to explore context and the decision-making process in schools because learning environments are changing as access to the Internet and other information technologies increases (OTA, 1995; Benton School, 1997). Implementation systems for acquiring, installing, and servicing information technologies need to be addressed, monitored, and revised with respect to the actual teaching and learning that they facilitate. As these systems become integrated into special education environments, Autonomous teacher decisions become further complicated in special education settings. For example, not only are there additional questions about school structure, but individual student needs and abilities as well, in particular where these needs dictate which instructional approach can be used (Rivera & Smith, 1997). For students with emotional and learning disabilities in more restrictive settings, there are added benefits when teachers work with psychologists, behavior therapists, and other related professionals to implement classroom instruction. However, classroom instruction can become a secondary issue when IEP teams are confronted with severe inappropriate behavior, even though instruction can contribute to or alleviate inappropriate behavior. Some even argue that the terms we use to classify and treat students with behavior problems places the blame on the student rather than "questioning the value of school structures and regimes" (Szasz, 1972) that may contribute to student failure. For students with learning disabilities, frequent experience with failure is common often leading to a subsequent "disenchantment" within educational settings. Providing opportunities for team documentation of combined successful teaching and behavioral practice are necessary, especially when new technologies or teaching strategies are identified that enhances student motivation, self-efficacy, and learning.

Documenting New Ideas and New Technologies

Issues currently exist with documenting student achievement, conducting appropriate assessments and continuing issues for identifying and accommodating students who have learning disabilities and associated behavioral and emotional disorders (Morrison & D'Incau, 2001). As technologies become available for instruction, teaching and learning continues to evolve. It has been generally difficult for researchers to document the influence of the Internet even though networking school environments is one of the most costly investments to be instituted in schools (OTA, 1995). Many of the educational benefits of information technologies may not fully be realized until such tools are integrated prudently into the curriculum (Dede, 1998; Norton & Gonzalez, 1998). This statement does not diminish the influence of new pedagogies and tools on teaching and learning; rather, educators need to continue looking at how classroom instruction has changed the response of educational systems to structure environments so that new tools and ideas can be implemented.

METHODS

Two questions were examined that focused on the integration of information technologies into a restrictive educational setting for students with emotional and learning disabilities:

(1) How does the special education school context influence teachers' use of new technology tools and the Internet?
(2) What are the systemic issues for supporting the adoption of an innovation in settings for students with learning and emotional disabilities?

Site selection and methods for data collection and analysis are discussed.

Site Selection

For the purposes of this study, research questions were addressed through an 8-month course developed through a university-school partnership as an ongoing training seminar for teachers working within the school organization. The course, EDSE600 – Technology Integration in the Classroom, was designed for teachers working with students having emotional and/or learning disabilities (ED/LD). Course content included emerging technologies, cognitive and constructivist learning theories, technology integration strategies, and the use of the Internet for teaching and learning.

To maintain confidentiality for this study, the education system is referred to as the Training and Development (T & D) foundation and schools. The Training and Development organization consists of four schools, one elementary/middle and three high schools. There are approximately 60 to 100 students in each of the schools. The T & D schools each serve students with the primary diagnosis of emotional disabilities. Many of the students have additional learning disabilities, attention deficit disorders with and without hyperactivity, and/or severe cognitive deficits. These schools operate under an umbrella organization called the Training and Development Foundation.

Five teachers from the course were selected based on their willingness to participate. The five teachers came from the three high schools within the Training and Development school. All of the teachers had a bachelor's degree and three individuals have advanced degrees. All of the teachers were first year teachers and working on initial state certification for ED/LD. Two of the teachers worked at Lee Training and Development School (Virginia), two of the teachers worked at Braddock Training and Development school (Maryland), and one teacher worked at Stevenson Training and Development school (Maryland). Each of these schools is a private non-profit educational entity that work with students having severe emotional disabilities and are affiliates of the "Training and Development" organization.

These teachers worked with students having a primary diagnosis for emotional disabilities. The students in these high schools ranged from age 14 to 21. Many of the students had additional developmental and learning disabilities, mild mental retardation, attention deficit disorder with and without hyperactivity, and/or speech and language issues. Many of the students had reading levels well below that of their comparative grade level peers.

Students are placed with the Training and Development schools after being unable to maintain a regular education or special education placement in their home school and/or school district. Although some of the students served by the Training and Development are mainstreamed back into their regular education setting, the majority of students complete their high school education with the Training and Development school. Each of the five participants completed two 1 and $\frac{1}{2}$ hour interviews. The first interview was completed mid-way through the course. The second interview was completed after the course was over. Two administrator interviews were completed mid-way through the course.

Data Analysis

Qualitative research is typically associated with a relativist worldview where meaning is contrived through situation and interaction with persons in different environments. Herbert Blumer (1969) described how social interactions influence individuals:

(1) People relate to things based on the meaning those things have for them,
(2) People learn the meaning of things in interaction, and
(3) People filter what they learn through their own perceptual lenses, which are based on their own particular socio-cultural and psychological backgrounds.

There is a wide degree of idiomatic research (research that specifies instances of specific technology influences that vary from school district to school district) available on regular education technology integration and teacher pedagogical beliefs. Most of the literature available on teacher innovations related to technology is relative in nature and describes how technology works with isolated school districts across the country making efforts to train teachers and implement technology within their schools (Southern Technology Council, 1997). A qualitative approach was used for this study due to an exploratory framework emphasizing process and development and an exploration of attitudes about teaching and learning with the Internet. Because of the changing nature of technology, studies need to include exploratory as well as fixed experimental measures to develop a complete understanding of technology related teaching and learning (Russek & Weinberg, 1993).

Data Analysis

A qualitative software tool called NU*DIST was used for managing the research data. Data from the interviews were imported into the *NUD*IST* program using *Dragon Naturally Speaking*, a continuous voice recognition software program that allows the user to input data to a computer without the use of a keyboard. NUD*IST facilitated the development of themes organized into nodes, text coding and searches, and organized research memos for each node and child of that node. NUD*IST reports of nodes were used to justify emerging themes and to develop results.

The analysis of data was on going throughout the study as themes emerged from the data. Data codes were inductively generated using a grounded approach of Glaser (1978) and were further developed during data analysis. Teachers were encouraged to comment on the themes identified during the interview process. The data from the second set of interviews was coded with themes based on the first and subsequent interviews. During the data analysis, negative cases were continuously explored crosscheck emerging themes. Teachers were provided many opportunities to provide examples of technology integration that did not work and were very forthcoming with information about their students, the school, and the overall system. The interview environment was established to allow teachers to describe how they integrated the Internet into their teaching practice. Interviews were coded immediately following the session and themes were identified. Another researcher was used to account for inter-rater reliability, where R = number of agreements/number of agreements + number of disagreements. After individual coding was conducted based on 7 selected themes, a match of 87% was found for similar coding processes. Member checks, peer debriefing, and triangulation were used to check for information about codes and emerging themes and to increase the accuracy within the data.

RESULTS

Systemic Adoption of Innovations

Teachers used the Internet for building transition and employment-related skills, augmenting reading and writing activities, and enhancing student motivation to complete classroom activities (see Castellani, 2000). Even though teachers realized innovative teaching practices in their respective classrooms, there were ongoing issues with how the T & D school philosophy and school setting influenced teachers as they attempted to integrate the Internet into their daily instruction. Administrators and related school personnel encouraged the use of the Internet for instruction. The system, however, discouraged teacher innovation by restricting teacher choice. Resources for teachers to integrate technology into the curriculum were difficult to obtain. The following section explores how specific school context issues influenced teachers' use of new technology tools, including the Internet, in their instruction. Additional facilitator and barriers to adoption of innovative teaching practices are explored. School context issues begin with the T & D school philosophy and school setting.

The Struggle Between Behavioral and Academic Intervention

At the time this study was conducted, the T & D school philosophy was based on a therapeutic framework and a behaviorist approach to teaching and learning. Throughout the course of this study, the integration of technology into the organizational philosophy of the T & D schools was under continued development. Administrative and teaching staff advocated for the use of technology in the classroom and struggled to define the role of the Internet in the therapeutic environment. Administrators, teachers, and staff seemed to constantly struggle with a balance between academics and behavior. The most ostensible reason is the structure of the institution as a psychiatric facility, with a "revamped focus" as an academic institute responsible for educating students with severe emotional disabilities. When the school was housed within psychiatric facilities in the immediate area, the emphasis was clearly on emotional and behavioral issues. After moving to less restrictive physical and philosophical setting, the mission of the T & D schools shifted from emotional to include both emotional and academic intervention. During the five years prior to this study, the school population tripled. Along with both teacher and student population growth, the size and presence of the T & D schools in the community grew accordingly.

In the Stevenson T & D school, there was a strict behavioral structure for students and teachers. Donna, the principal for the Stevenson school and the primary motivator for the EDSE600 course, describes her philosophy:

> John: Do have a certain point of view in the way that you deal with students, I mean could you classify yourself in any way?
> Donna: I would, very clearly, and I think that every principal is different, but I am a real strong disciplinarian, regardless of what clinical issues and behavioral issues are presenting themselves. I think that particularly with this population . . . we have really strict rules about cursing, threatening . . . you're just not allowed, and a lot of the kids will say that's just how I am, that's why I have emotional problems, and I will say no that is not why you're here. You're not here because you curse. You are here because that is how you manage your anger. We are going to teach you another way to manage your anger, so I am a very clear and firm disciplinarian. I think you hold fast to clear expectations . . . which are what these kids don't have when they came into a program like this. You set that, you're respectful to others, consequences are consistent, and then you provide rewards . . . it's not punitive. In fact, what we do is provide them with security and a firm approach to behavior with standard consequences and standard rewards, so you know what to expect . . . . That is my philosophy.

The T & D schools were responsible for providing all of their students with individualized education in the least restrictive manner. Since students came to the T & D Schools with identified emotional problems, there were conflicting issues about providing services within these facilities and the overall definition of "academic instruction." In many cases, the word "academics" was defined as another venue to work with behavioral and emotional issues. When considering a hierarchy of training needs, administrators and staff were consistently focusing on behavior as a precursor to academic learning and dealing with students' primary emotional and behavioral disabilities. As teachers discovered new ideas for teaching and learning that enhanced student motivation, the overarching system discouraged the integration of these ideas into student programming. Teachers created a more expressive curriculum environment that allowed students to engage in open-ended technology and Internet-based activities. These ideas were both difficult to communicate and document. In daily behavioral review meetings, teachers were indirectly discouraged from offering strategies outside of the available behavior management system.

During an interview with Ann, the Education Director at the Stevenson T & D school, she was asked how the Internet was being used for instruction and about teaching and learning with the Internet in a purely "academic" sense and paired with specific content learning goals. Her response identifies some confusion about how academics and behavioral intervention needed to be paired:

> We offer groups that we're working on self-esteem depending upon what the students need, and there are a lot of components to this. We need to get "hands on" strategies that can help us integrate (academic and behavioral components). As we expand the academic, we will also need to do the same for the behavior. As of now we have not had groups go in the computer lab for therapy, maybe a couple of times or so, but for the most part, they have not gone into the lab as a group during therapy to work on their modules.

Ethan and Donna explained why the T & D schools were struggling with academics and behavior. Less than five to six years ago, the schools were based within psychiatric facilities and have just recently become "schools." As they have moved to a less restrictive environment, the schools were dealing with their identity.

> Ethan: I think you have to remember that this place started only recently, and has only been around for about five years, so they are still going through trying to find their identity as a school, as opposed to being a psychiatric institute.

Donna acknowledged these ideas about the T & D schools. She expanded on these ideas and further explained the issues confronting the school in terms of academic intervention, mainly because they have focused on getting the system and student behavior under control. She added that as behavioral influences decreased and were "under control", they would be in a position to move forward on academic issues. Since the Stevenson T & D school was "on top of" the emotional issues, academic issues could be emphasized:

> Donna: One of the things that actually relates to staff development that has struck me recently is that the program is in pretty good shape. [The program] has its lows and highs, but once you've got the formula you can pretty much get there. But during these times when it is so quiet, and actually sometimes boring for staff, those are the times where you need to be creative, and need to be coming up with new and more creative instructional programs, pieces of the curriculum . . . so we're really ripe for that. You know, we need to know where to go now that we've gotten to this point. We don't deal with other stuff now, we are beyond that, and we are at a place for really thinking we need to bring technology in that can help these kids even more.

The 9-month EDSE600 course began in the month of November and continued until June. When the course began, all schools were connected to the Internet. At the beginning of the school year, each of the T & D schools was in a position to integrate new ides into the curriculum. The schools were equipped with new technology labs, which consisted of "very expensive, state-of-the-art computers." Donna was aware that many staff did not know how or simply were not using them. She was concerned about students' academic instruction and further stated that, "if the students do not use the computers, they're not going to be ready for the job market."

Teachers attempted to integrate technology into their curriculum. In many cases, student behavior was a result of ill-conceived lessons combined with individual student learning differences. The following conversation was recorded on an electronic learning community associated with the EDSE600 course. This conversation is one example of how teachers used technology to deal with both behavior and student learning. Maggie explains a difficulty she was having with one particular student:

> I have a student who consistently disrupts the class, seeks attention, and does not get the work done that is required in the course. Her *reading* level isn't that high, but she can *write* around a 4th grade level, even though she is in 10th grade. I am not sure if she can do the required work.

Teresa responded to Maggie about her disruptive student:

> You say you have no idea how capable the student is of doing the work or if he has a specific learning disability (SLD). It seems to me that it is crucial to find out whether or not the student is capable of doing the assigned work. As you know, a student is most likely to be disruptive when he cannot do the assigned work or when he finds it exceedingly easy. Access in the classroom to a computer with English software programs would be helpful. I have been able to motivate bright disruptive students to complete class work and then use appropriate software to increase their skill level when the class work is completed.

In the next entry, Ethan explained how he combined instructional and behavioral intervention through the use of available technology:

> I have found that Internet access is a great motivator for disruptive students to become more focused. If you wanted to help her, she could import film audio and video bytes into her presentation. Finally, upon completion of the two tasks she would earn access to areas of the Internet that centered about her interests, with the stipulation that she would create an English paper that addressed those interests.

Philosophically, the T & D schools have a mission to work with students on academic and emotional issues. Although Donna has a clinical approach to dealing with students and the school environment, she was "tuned into" the need for increased academic instruction; however, she did not know what that would include. Ann was interested in pushing the technology to increase academic instruction; however, since she had limited skills, she was not at a point to model appropriate integration strategies for teachers. And finally, there was a tension between providing opportunities through technology and limiting students through the use of behaviorist intervention strategies.

Opportunities for documenting these on the IEP or sharing with non-teaching staff were not common. Teachers had an opportunity to discuss these issues through the electronic discussion board because they were taking the class. Teachers consistently argued that these discussions should be included in their daily behavioral "wrap-up" meetings with counselors and administrators. As teachers discovered new ideas for teaching and learning that enhanced student motivation, the process discouraged the integration of these ideas into behavior management planning. Many of these ideas allowed students to engage in open-ended Internet activities and work with self-selected reading and writing materials. Although students achieved academic gains and remained on task for longer amounts of time, these strategies were difficult to document given the current IEP, behavior management system, or school constraints on available teacher time for this task.

Appropriate Documentation and Transference of New Knowledge

Information was gathered from both teachers and administrators about the IEP to see how they were following the IEP in "spirit" and not legal requirements. The schools had recently completed a program review by the state and were found in compliance with the law. However, teachers questioned the role of the IEP and its influence on instruction. Student IEPs addressed mainly *behavioral* systems and broad *academic* goals. Teachers noted that the IEP did not include technology. The primary emphasis for *learning* objectives within the T & D school environments were behavioral:

> Donna: It is interesting when you are dealing with this population. They are intact cognitively and once you get on top of the behavior and clinical emotional issues, they are very functional . . . you get on grade for math and you get on grade level for reading. The IEP is useful, but for non-LD kids, the real focus is more on the clinical issues.
> John: So I would imagine that many of the goals on the IEP would wind up being behavioral goals?
> Donna: Well, regardless of the disorder, emotional disabilities tend to manifest themselves in adolescence in a variety of ways. They have problems taking directions, they have problems with authority . . . it is hard to individualize unless you're actually beginning to uncover some of the clinical issues that are causing the problems and then you can do it.

The teachers responded that the IEP did not seem to fit well with classroom activities and or contribute significantly to the wide range of individual student needs. When goals were developed for academic instruction, they were broadly defined, so teachers were not given a clear indication of what they were supposed to be doing with students in the classroom. As students entered the school environment, they came with an accompanying IEP. Because of the location of the schools, teachers needed to respond to IEP requirements and formats from three different states and eight different school districts depending on the surrounding county. First, Maggie reports:

> It has pretty much been left up to each individual teacher to go through each IEP . . . the main concern this year is that I haven't had the time to look through all 53 of them and say 'Oh well, at this point in whatever lesson, I am going to have to worry about this for that student' . . . I don't know what teachers really have to do with it, umm, is helpful to know, but I think at the same time I did not write the IEP, someone else wrote the IEP from another class or another experience with that student, and is not necessarily the same thing that is going on in my class.

Second, Steve offered his thoughts about the IEP and how he thought it actually influenced his classroom teaching:

> I sat in on an IEP, as teachers we get together and write down what we think they can master and not master. I really and honestly think, and I'm still dealing the whole issue, the way IEP is set up now is just to satisfy funding. Okay, a student has mastered that and now we can move on. The writing on the IEP is difficult to implement for these students and I think the words "mastering and not mastering" are just too concrete. And to try to make your lesson plans based on the IEP is pretty much impossible when you look at it. We have a thing on our behavior point sheet, "IEP goal today". One student's goal is to identify anger and frustration and to learn how to deal with it. I am like, what the hell is that? This kid is on a third grade reading level and he is supposed to understand that? And there are all kinds of stuff like that. I think we can do a lot more with less. Here you go . . . this is the IEP for one kid, 10 pages, it doesn't really make any sense and doesn't respond to what I am doing in the classroom . . . to really break it down, I think it is pointless.

Like Steve and Maggie, Warren commented that the IEP was for something other than classroom instruction. The roles of teachers and the development of the IEP were not clearly defined.

> Warren: Some kids claimed to be transitioning back into public school, but they really don't . . . I guess the IEP comes into play when they are transitioned back into the public school, and that is what the social worker or psychologist and team leaders look at. For teachers, the ability to use the IEP as a vehicle for instructional guidance was diminished as a result.

The IEP is a central student document that is designed to enhance instruction; however, teachers did not feel that it was useful for their classroom. The IEP was not responsive to daily student-teacher interactions and did not include technology. In several cases, teachers had difficulty following what was recorded on the IEP, mainly because it was seen as something developed outside of the context of the 'day-to-day' class activities. When it did address student issues, the goals were mainly behavioral. In terms of technology, none of the teachers identified a technology goal or objective, yet they were being asked to integrate technology into their instruction through a course in technology integration. Teachers stated that because of competing priorities, they spent the majority of time marking their behavior sheets and record broad objectives and IEP progress. As a result, many teaching innovations with technology were left undocumented on students IEPs. Given the high turnover rate for teachers and attendance issues with students, what worked for classroom instruction was not passed on from teacher to teacher, classroom to classroom, and school to mainstream environment.

Teacher Responsibilities and Individual Choice

Teachers agreed that school personnel should focus on behavioral and emotional issues, but it seemed difficult for teachers to continually deal with behavior when they were hired to work primarily with academics. In many cases, the T & D System restricted teacher choice by deciding student groupings, dictating curriculum, instituting a school-wide behavior management system, and allocating resources for technology in the schools. Teachers had to negotiate resources, the computer lab, and find methods for communicating their teaching practices within the given behavior management system.

School-wide Behavior Management System

Most teachers were frustrated with having to act like counselors rather than educators. Teachers were not required to restrain or intervene when students were "out of control." Assistants, counselors, administrators, and social workers were required to deal with these issues even though much of the "teaching time" was focused on behavioral intervention. Teachers often seemed confused about their role in the school.

> Ethan: When I worked in the New York system, we did the counseling, and we restrained the kids and really dealt with behavior management. There wasn't a behavior management team, there was a school psychologist who came in and did his thing, but outside of that we did everything, so when that is the case, there are really very little politics and not much room for disagreement. And there were not as many problems. Here, not that it is a problem, but you have different purposes in each area, and that tends to divide rather than finding a common ground. I accepted the job here knowing that I was hired only to teach. However, if I had accepted the job here with the expectation that I might have to restrain kids, I would have accepted that, actually.

When the school hired Ethan, he was told that he would not have to restrain students since there was a system in place to deal with students' emotional disabilities, such as trained social workers and a school psychologist. His primary responsibility was instruction and academics. However, much of his time was spent dealing with emotions, so this message was inconsistent. He did not really feel like he could separate these two issues. Teresa also explains this issue:

> Teresa: I have to be more detached about what is going on outside and more focused on trying to help the students, academically only. I get too much involved with a behavior here. I am not trying to be a counselor.

Steve, Warren, and Maggie also gave detailed examples of having to deal with student behavior and subsequent compromises in academic instruction. Steve and Warren expressed similar experiences at the Lee T & D school, although when dealing with behavior, the administrator did not take into account the influence on academics.

> Steve: Sometimes I think behavioral issues are put ahead of academic issues. I think my background of working as an assistant helps me to set limits with the kids. We seem to focus so much on them. Some kids came in and killed all the fish in the fish tank and they were put out for four days, and there was a lot of back and forth and I was working on job placements for them. I mean, if you try to take a kid out of therapy to finish up an assignment the counselors scream bloody murder. If they want to take the kids out of class to go talk about an issue, that is all right. So there are mixed messages.

Steve did not say that the students should not be disciplined, however, his frustration was over the influence that this had on a student's job placement, "which takes weeks of work," only to have it "wiped out" because a student was sent home for killing fish in a tank.

> Warren: [The school] where I work is so focused on psychology and psychotherapy that the education piece is a stepchild. That bothers me. That really bothers me. We get so meshed into people's issues with psychoanalyst "wannabe's," and when I saw it here, I said, 'we are not talking about learning, we are not talking about reaching these kids in the classroom, all we are talking about is behavior. And all we're trying to correct is behaviors.

The implications for this study was that the teachers complained about "not having time" to implement ideas in the course, even though the course had been arranged by the school administrator to support instruction. Warren, Steve, and Ethan consistently complained about not having time to work in the computer lab after school. It seemed like the issue of time was directly related to competing activities rather than their inability during the "normal" teaching day to work with new ideas and tools. Teachers did not advocate that behavior be excluded from intervention. They advocated for more attention on academic intervention, especially for incorporating ideas about new technologies. The common complaint was that if they were hired to teach, why were they not given time to develop teaching ideas?

Emotional issues are never completely resolved. In many cases, emotional intervention conflicted with the academic activities of teachers. Behavioral intervention was a necessary component to the school. However, the struggle between behavior and academics offered a unique tension to the T & D school environment, including how students should be disciplined and what constitutes punishment. At the end of each school day, teachers, administrators, and counselors get together in clinical meetings to talk about student issues. In this sense, the teacher was not removed from emotional intervention. In fact, this encouraged their involvement in behavioral intervention and, according to teachers, took away from "individual planning time and grading." These clinical meetings often discouraged teachers from preparing for the EDSE600 course, communicating with one another about what technology was working, or time to learn new skills on the computer. This directly influenced teachers' use of the Internet in the classroom because if teachers did not feel comfortable with the computer, they would not use it during classroom time for fear of disruptions, difficulty, and additional stress while teaching.

Negotiating The Computer Lab

The administrative team decided to set up computer labs around the schools rather than placing computers in classrooms. The computer lab was an ongoing issue throughout the study. The teachers identified issues with technology use that are not exclusive to special education. However, due to the nature of resources in general and the structure of the school setting, teachers' complaints about the computer lab were slightly more significant because of the nature of the "special" school setting. For example, teachers in a regular school would not be displaced because a third of the students were removed from their regular classroom settings because of such a "lock-down." However, Donna explains how the decision were made to develop the computer lab:

> John: In terms of why the computers were set up in the computer lab, do you have the reason why you put the computers in the lab?
> Donna: (Long pause) A couple of reasons I know, we wanted to be able to teach technology to classes, and we very much also wanted to be part of vocational education, so we wanted to have it into classroom, as really the focus of the vocational teacher's classroom. They teach about three periods, so that is how other classes are able to use the lab . . . but it was sort of organized to accommodate the Voc-Ed component.

Although there were benefits to having the computer lab, there were several issues that resulted. At the Stevenson T & D school, the lab was relatively inaccessible for approximately four months:

> Ethan: I have been in this luxury situation having the computer lab because of the shutdown, of course I have the computer class that has always been in there, but since the shutdown, I have had the computer lab all to myself.

At the Braddock T & D school, there were several times when the computer lab was closed because of students' inappropriate behavior while working on the computers:

> John: You had mentioned the philosophy of using the Internet; who came up with the philosophy, or how was the philosophy decided?
> Maggie: Actually, the week I was out, when they had started using the Internet, something happened with a few students so they closed the computer lab.

At the Lee T & D school, Warren stated that having the computer lab made them feel uncomfortable about working "outside of his own classroom," especially since he could not control "all of the students at the same time." Steve did not have this trouble, but did comment that the computer lab sometimes required additional teaching staff. In these cases, the use of a computer in the lab rather than computers in the classroom prevented students who were acting appropriately from using it during the week. It also seemed to place additional pressures on the individual teachers required to operate the lab.

The teachers often expressed that they would like computers in their classrooms. However, none of these teachers said that they did not like to use the lab because it was a computer lab. Their preference was for a computer in their classroom but they saw the advantages to a computer lab as well:

> Ethan: We have to get to the point where you have access to computers when you need them. So whether this is one or two computers in the classroom, or placement of the computers in a lab, it doesn't matter. Eventually, it will get to the point that there are enough computers for each class and also a computer lab.

Availability of professional resources for academic instruction.

There seemed to be an overall level of concern about the lack of professional resources available to the classroom teacher to implement instructional activities. Although texts, art supplies, and gym materials were present, teachers complained that the breadth of materials necessary to work with students having such a wide range of needs were not adequate. The lack of resources was mainly a result of the separate nature of the school setting from the regular education environment where access to general education materials is greater.

> Ethan: There are a lot of concerns from teachers about the availability of things. I am not only talking about computers and books, but the physical structure itself. You know if you look at the facility, there is definitely a need for more space.

Apart from the Lee T & D school, the Braddock and Stevenson schools were within relatively small spaces. There was adequate room for general daily practices, but there was no room for growth. Also, teachers did not have access to a library within their individual schools. The T & D schools designated a library within each school, but these libraries are not adequate to meet the needs of the teachers and students:

> Maggie: Look at the resources that I have available to me. You can look around here and see our library, I keep telling them that we should just stop the lying and call it what it is . . . a conference room.

Warren was irritated about the lack of resources. He stated, "the school spent so much money on the Internet and I still don't have any books." The time to develop new resources was lacking. He thought that the emphasis on behavior was one reason for the paucity of professional resources.

> Warren: We are not talking about learning, we are not talking about reaching these kids in the classroom, and all we are talking about is behavior. And all we're trying to correct is behaviors, issues, we're not doing anything . . . even though I give them homework, the parents call and say they are not getting any homework. The administration was quick to tell me to give the student homework. Well, six months have gone by since the beginning of the school year and we still don't have any books. You know, that really troubles me a great deal.

When I visited the schools, the available resources seemed to be the same, including software and hardware, and the available library. I also was not aware of a music room; however, the schools did have art classes. In terms of the gymnasium, the Lee T & D school had the largest facility. At the Braddock and Stevenson T & D schools, the indoor and outdoor space was limited:

> Ethan: I had more resources in the New York system. I have a few resources for PE here, you can see outside of the window what I have available. I mean it is a little bit more than that patch of grass that you can see, but not much.

Teacher choice was often influenced by the available resources for instruction, barriers to using the computer lab, and by the struggle between emotional intervention and classroom learning. Teachers felt that their ability to work with innovative forms of teaching and learning were indirectly discouraged, even though the administration was vocally advocating for technology use in the classroom. The following section is a discussion of these issues with emphasis on their impact within the teaching and learning environment. Additional student access to technology is described outside of the school community. Identification and implications for these students as members of an increasing digital divide are discussed.

DISCUSSION

It is necessary to explore the distinction between academic instruction and behavioral intervention, given an environment established for working with students having emotional and learning disabilities. The environment, or therapeutic milieu, directly affected the role of teachers by requiring teachers to conduct themselves in a consistent manner within a specific behavioral plan for intervention. When considering the Internet as a tool for the T & D school environment, it is important to discuss the impact of this new tool on the academic and behavioral intervention strategies already in place to work with students.

When Ann and Donna stated that they were "in a position" to work with students on academic issues when behavior is under control, they clearly meant that they cannot work with academics when student emotional issues were out of control. Psychiatric institutions generally focus on emotional, behavioral, and therapy issues. As the institutes have moved from hospitals to academic learning centers, teaching and learning has improved. However, teachers' and administrators' describe the T & D school environment as a psychiatric institute. The research suggests that identifying the academic variables involved with students having behavioral disorders might greatly enhance the ability for special education programs to become more efficient and effective (Fessler, Rosenberg & Rosenberg, 1991). In addition, identification and treatment of students with behavioral disorders is primary even though adequate identification of dually paired learning disabilities and issues surrounding intervention are often left unresolved (Glassberg, Hooper & Mattison, 1999; Kauffman, Cullinan & Epstein, 1987).

One goal of the "therapeutic environment" is to place students with varying emotional disabilities to balance the demands on the teacher and to pair students by opposing emotional states. For example, Donna expressed that they will often pair students in a state of severe depression with students who are manic to bring students with depression "out of their shell" and to balance emotional difficulties. In the classroom, however, this places additional demands on the teacher since students are being grouped for emotional intervention rather than academic ability levels. A teacher's response is to provide an appropriate balance of structure to accommodate student learning.

The use of the Internet and the computer lab seemed to encourage a more student-centered approach in the T & D schools. In many student-centered approaches, the individual is in control of his/her learning. Rather than establishing strict classroom structure, teachers working within a more holistic environment are responsible for providing support for learning. Holistic environments are different from traditional special education environments where instruction is developed, controlled, and delivered by the classroom teacher (Rivera & Smith, 1997). Even though both approaches may be beneficial to teaching and learning, there is an inherent tension between these two structures. In order to set up a more holistic environment, some level of control needs is relinquished to the student. It is often common for special educators to direct instruction based on a need to remediate the present lack of skills. As students advance to higher grades, intervention efforts for community, real world, social and academic interactions are more compelling.

The presence of student deficits combined with frustration over learning encouraged more teachers' use of more open-ended classroom experiences. For Steve, Warren, and Maggie, student-oriented learning was the primary mode of instruction. The goal of teachers seemed centered on real world and individualized experiences. Teachers felt students needed access to community skills to increase their ability to function in school, home, and community settings.

Within the T & D schools, students were grouped by emotional disability rather than cognitive disability. In fact, both administrators, Donna and Ann, hesitated to admit that cognitive issues were even present in the students. Rather, the focus was on dealing with students' primary emotional disabilities. Teachers and administrators held different opinions about grouping students. In her final journal entry for the EDSE600 class, Maggie came to a realization about the integration of emotional and cognitive disabilities and student groups.

> Maggie: A seemingly obvious thought occurred to me the other day. I've been wondering what I'm doing wrong with my lesson plans that students aren't staying engaged and in class. Beyond the whole issue of their emotional disability, I've been wondering why lessons aren't going as smoothly as I would expect. What occurred to me is that some of the students have learning disabilities and are diagnosed as ADD or ADHD, and some are merely classified as emotionally disturbed. Therefore, a lesson that would engage or work for a student diagnosed as ADD would not work for someone who only has a learning disability. . . . What I think would work better than grouping students by age, grade level

> or emotional disability would be to group them by learning style. If you are a visual learner but have dyslexia, then you do not need the same lesson as your friend who has been diagnosed as having ADHD.
>
> Having just come out of college, I have heard a lot about community learning and having students paired up in class or in mixed groups where there is a higher-level student who can instruct and model for his/her partner or the others in the group. It seems to me that people learn by mimicking adults and those around them anyway, groups made up of students of varying levels would work. I hope to be able to bring this suggestion to my education director when we make up the specialized learning groups for our school in the fall. It might make lesson plans easier to write classroom management easier to control and could even increase the amount of learning going on in each classroom.

In this environment, it is assumed that the appropriate student mix will encourage student emotional growth. However, teachers expressed difficulty with competing priorities for their time in the classroom. Teachers questioned their role in the school. Should they work on academic instruction or deal with students' continual emotional and behavioral needs? If they were to work on academic issues, what is the priority for considering student groupings based on instruction rather than emotional intervention and was there an appropriate balance?

In many cases, teachers felt they could achieve an appropriate balance in the classroom if administrators and counselors provided more support for academic instruction. They seemed to achieve a more balanced approach to learning in the computer lab and with the Internet, because students were on task, involved in learning, and therefore the need for behavioral intervention was diminished. Learning was more individualized, so that teachers' concerns about student groupings and classroom management were also diminished.

This study also found issues with utilizing Internet based tools in the classroom. Often times, students were not allowed to interact through e-mail and electronic communication. Although there was a school emphasis on enhancing appropriate communication, teachers and students were not encouraged to interact with others outside of the school setting. There was a fear among teachers and administrators that these students would act in inappropriate ways. At one point, the administration disallowed communication with outside resources. However, during the summer months, teachers and administrators were beginning to feel more comfortable with the idea that students could interact with outside individuals. In fact, this interaction could provide an opportunity for students to work on appropriate communication in a controlled context where teachers have the opportunity to monitor student e-mail and asynchronous and synchronous discussions.

Teachers disagreed with many decisions made by counselors and administrators on a daily basis. However, one consistent area of agreement was the feasibility of the Internet for instruction. Teachers and administrators agreed that the Internet was an emerging tool available for classroom instruction, but were cautious about using the Internet with students having such complex disabilities. Before jumping into an activity, teachers tended to consider the relationship between what they were teaching and how they thought students were going to react. This prevented them from using e-mail and possibly other available tools as well.

In many school environments, inadequate training and the lack of opportunity or incentives are primary forces that impede the efforts of technology integration in the classroom (Wager, 1997). In many technology oriented teacher-training programs, the focus is on improving the skills of teachers. Within the T & D setting, teachers needed more time outside of the course to plan activities integrating technology for their students. The immediacy for meeting student needs was enhanced as students moved up grade levels and became increasingly at-risk for school dropout and/or general school failure. As a result, teachers needed time to develop motivational and academic strategies that would capture student interest and enhance teaching and learning. Since students were generally motivated to work with technology and the Internet, emphasis for future training should be placed on developing teachers to use such tools and strategies.

Darling-Hammond has indicated that school reform is successful and sustainable when administrators, counselors, and teachers share a common vision, build a collaborative culture, and weave changes into the organization of the school (Fullan, 1991; Ruddick, 1991). Senge (1991) comments that within learning communities the success or failure of one person influences others' successes or failures. Individuals across various levels of education, who are technology literate, are working towards the development of innovations that are necessary for successful technology integration. However, administrators with technology expertise or ability are not in every school. Unlike other studies (Alberta, 1993; McKinsey & Company, 1997), the T & D schools were supportive of technology integration and were willing to provide teachers with necessary technology training.

As school personnel were defining the functional applications of the Internet, their disposition for technology changed. During the course of this study, the institution operated within a continuum that both encouraged and discouraged the use of the Internet for classroom instruction. This study found that the available time for effective planning was influenced significantly by competing teacher priorities. This study also found that there was a distinct struggle between behavioral intervention and academic learning. Results indicate that teacher training for technology integration in the special education environments may not be effective if all stakeholders are not included in technology development and training.

STUDENTS, COMMUNITIES, AND THE DIGITAL DIVIDE

Teachers expressed that community influence was a concern for two main reasons. First, they typically did not interact with parents academically because the majority of contact with parents was behavioral. The teacher contact with parents was marginal. When contact was made with parents, it was initiated based on emotional or behavioral issues. Teachers expressed that many students had disruptive, abusive, and isolated home environments. In many ways, teachers were dealing with a widening group of students in an identified digital divide area. Second, teachers were very concerned about supporting students with technology and providing them with access to information technology. In addition to poverty, these students exhibited behavioral problems and had difficulties with drug addiction and crime. As a result, teachers felt compelled to provide students access to different resources at school due to a paucity of these in the home, especially technology. Teachers felt that students would fall into a gap that would prevent them from being able to independently functioning in an information-based technologically rich society.

Teachers expressed that "providing opportunities for students who do not have computers at home was a very important reason for using the computer in school." Issues of economy, environment, and society were consistently represented in teacher interviews. Ethan reported that "students do not have computers at home and they need the skills if they are going to compete in the job market." Steve said that "only two or three of his students had access at home" and therefore did not have any of the skills needed to work with computers in class. Lastly, Teresa stressed that students were so interested in computers in class because "they do not have access to them at home." In addition to teaching and learning, the T & D system struggled with a unique "digital divide", influenced by parents and community settings and further complicated by disability, drug addiction, and violence.

Teachers had very limited access to parents of these students on an academic level. When parents were involved, it was typically around problem behaviors in the school. Maggie noted that once she called a parent to say what a wonderful job her child had done in school that day and the parent was "in shock". Maggie stated often that parents did not express "even a curiosity factor." Maggie said that parents often thought it would be nice to "hear something positive when we come in for a meeting" rather than always having to deal with behavioral issues:"

> Maggie: I guess with the population that we are working with . . . there are those things going on, but you have limited time in these meetings and you need to address the severity of issues. So when they call, it is usually in response to a therapist or teacher phone call.

Steve also expressed that his contact with parents was limited, even though he was involved with transition planning and community placement:

> Interviewer: Have you had any parent input?
> Steve: Umm . . . for me, no. I have very limited parental contact unless it has to do with jobs or something like that, and then I do.

Teresa noted that during an IEP meeting, she took the parents around the school. They were "surprised to see the computer lab" and responded to her that they were happy the school "was working on academics too." Teachers were working with students on technology-related skills, students remained on task for longer periods of time, and the learning environment was incrementally improving. However, systems for sustaining such activities outside of the classroom were difficult to determine. Teachers felt that internal and external forces continually influenced their classroom teaching practice in both positive and negative ways.

CONCLUSIONS

The purpose of this chapter was to explain how a restrictive educational facility attempted to integrate the use of the Internet as a tool for literacy instruction into their schools through the view of the teachers and administrators involved in the implementation process. The outcomes of this research suggest that:

(1) Assistive and instructional technologies are imperative tools that can accommodate students with emotional and learning disabilities,
(2) Documentation on advances in classroom instruction around emerging technologies needs to reflect appropriate practice and involve all stakeholders,
(3) Students with emotional and learning disabilities are increasingly at-risk for becoming part of the gap of those who do not have access to technology, and
(4) Systems implementing technology innovations for students with learning and emotional disabilities need to consider decision-making processes and responsibilities for systemic change in therapeutic technology-based interventions.

This study found the T & D setting to be an integrated system involving many individuals advocating for the instructional needs of each child. The level of service and individuals involved in interventions often increased as the number and type of student variables increased. The number of individuals involved in making these decisions provided increased challenges for adopting new technology strategies. Teachers working in the T & D schools repeatedly expressed that the time provided for training was not adequate to support the integration of technology into the school curriculum. The priority within the T & D schools for using the Internet was high. The amount of time necessary to integrate new tools into the curriculum did not seem sufficient due to competing priorities on teacher time. Certification, re-certification, IEP development, parent contact, and the amount of time devoted to clinical meetings and discussing student emotional needs appeared to be overwhelming for teachers. Since many of these individuals were within their first-year of teaching, the presence of competing priorities resulted in frustration and added to teacher burnout.

This study found that school administrators did not have an established plan for technology integration. They were operating under a medical model that dealt with students' primary emotional disability and which focused on behavioral rather than learning goals.

The dialogue between teachers, administrators, and counselors was behaviorally and emotionally oriented. Secondary issues, such as ADD, ADHD, or learning disabilities were left relatively unexplored by school staff. However, the presence of the Internet began to change the discourse of teachers, administrators, and counselors within the T & D schools.

The EDSE600 course was an administrative effort to provide teachers with technology training. Even though teachers raised concerns over the amount of time given to technology training, support was available through the school. Teachers expressed additional opportunities beyond the EDSE600 course to take courses for certification or general skill building. The administration effectively supported teachers through the allocation of resources and the training necessary to implement them. Although the training may not have met all of these teachers' certification and training needs, the priority for technology teacher training was present.

The use of the Internet did not significantly alter a school approach to teaching and learning. A study of long-term teacher growth, student access to technology, and school change would be necessary to determine if the presence of technology was an answer to even some of the problems the T & D schools are facing. It is possible to determine the influences that instructional technologies have for accommodating individuals with disabilities and how schools can fulfill their legal requirements for considering assistive technology for individuals with emotional and learning disabilities. Given the high turnover rate of teachers in restrictive special education environments, the sustainability of technology practices may necessitate a system of ongoing training and development for all individuals involved. Otherwise, lasting training effects will disappear with changing personnel. Supporting teachers working within complex systems and increasing student access to technology may naturally prove to accommodate students with disabilities. In some cases, it may even serve to reduce school failure.

REFERENCES

Behrmann, M. M. (1998). Assistive technology for young children in special education. In: C. Dede (Ed.), *ASCD Yearbook 1998: Learning with Technology.* Alexandria, VA: ASCD.

Benton School (1997). *What's going on? The learning connection, schools in the information age.* Washington, D.C.: Benton Foundation.

Brandt, R., & Perkins, D. (2000). The evolving science of learning. In: R. Brandt (Ed.), *ASCD Yearbook 2000: Education in a New Era.* Alexandria, VA: ASCD.

Center for Applied Special Technology (CAST) (1996). *The role of online communication in schools: A national study.* (Available: http://www.cast.org).

Castellani, J., & Jeffs, T (2001). Reading and writing: Emerging teaching strategies, available technology tools, and the Internet. *Exceptional Children,* (3), 45–59.

Castellani, J. D. (2000). Strategies for integrating the Internet into classrooms for high school students with emotional and learning disabilities. *Intervention in School and Clinic, 35*(5), 297–305.

The Cognition and Technology Group at Vanderbilt University. (1994). Multimedia environments for developing literacy in At-Risk students. In: B. Means (Ed), *Technology and Education Reform.* San Francisco, CA: Little, Jossey-Bass Publishers.

Cuban, L. (1979). Determinants of curriculum change and stability, 1870–1970. In: J. Schaffarzick & G. Styles (Eds), *Value Conflicts and Curriculum Issues.* Berkeley, CA: McCutchan.

Cuban, L. (1993). *How teachers taught: Constancy and change in American classrooms, 1890–1990* (2nd ed.). NY: Teachers College Press.

Dede, C. (Ed.) (1999). The scaling-up process for technology based educational innovations. In: C. Dede (Ed.), *ASCD Yearbook 1998: Learning with Technology.* Alexandria, VA: ASCD.

Fitzgerald, G. E., & Koury, K. A. (1996). Empirical advances in technology-assisted Instruction for students with mild and moderate disabilities. *Journal of Research on Computing in Education, 28*(4), 526–553.

Fitzgerald, G. (1990). *Using the computer with students with emotional and behavioral disorders.* (ERIC Document Reproduction Service No. ED 339 155).

Fullan, M. G. (1991). *The new meaning of educational change* (2nd ed.). New York, NY: Teachers College Press.

Glassberg, L. A., Hooper, S. R., & Mattison, R. E. (1999). Prevalence of learning disabilities at enrollment in special education students with behavioral disorders. *Behavioral Disorders, 25*(1), 9–21.

Hasselbring, T. (1994). Using media for developing mental models and anchoring instruction. *American Annals of the Deaf, 139,* 36–44.

Hasselbring, T. S. (1989). Making knowledge meaningful: Applications of hypermedia. *Journal of Special Education Technology, 10*(2), 61–72.

Higgins, K., & Boone, R. (1997). *Technology for students with learning disabilities: Educational applications*. Austin, TX: PRO-ED, Inc.

Individuals with Disabilities Education Act of 1997, Pub. L. No. 105–17 (1997).

International Society for Technology in Education (1997). *Technology standards for Teachers (On-line)*. Available: http://www.iste.org/specproj/standards/standard.htm (1998, February 20).

Holzberg, C. (1995). What works? technology in special education. *Technology and Learning*, 18–23.

Inman, L. A., Knox-Quin, C., & Horney, M. A. (1997). Computer-based study strategies for students with learning disabilities: Individual differences associated with adoption level. In: K. Higgins & R. Boone (Eds), *Technology for Students with Learning Disabilities: Educational Applications*. Austin, TX: PRO-ED, Inc.

Jacobson, M. J., & Spiro, R. J. (1995). Hypertext learning environments, cognitive flexibility, and the transfer of complex knowledge. *Journal of Educational Computing Research, 12*, 301–333.

Julnes, R. E., & Brown, S. E. (1993). The legal mandate to provide assistive technology in special education programming. *West's Education Law Quarterly, 2*(4), 552–563.

Kauffman, J. M., Cullinan, D., & Epstein, M. H. (1987). Characteristics of students placed in special programs for the seriously emotionally disturbed. *Behavioral Disorders, 12*, 175–184.

Lewis, R. (1993). *Special education technology: Classroom applications*. Pacific Grove, CA: Brooks/Cole Publishing Co.

MacArthur, C. (1997). Using technology to enhance the writing processes of students with learning disabilities. In: K. Higgins & R. Boone (Eds), *Technology for Students with Learning Disabilities: Educational Applications*. Austin, TX: PRO-ED, Inc.

Male, M. (1997). *Technology for inclusion: Meeting the special needs of all students* (3rd ed.). Boston, MA: Allyn & Bacon.

Mastropieri, M. A., & Scruggs, T. E. (1994). Text-based vs. activities-oriented science curriculum: Implications for students with disabilities. *Remedial and Special Education, 15*, 72–85.

National Center for Educational Statistics (1999). Statistical compendia: Education at a glance (On-line). Available at: http://nces.ed.gov/pubsearch/index.html (2, April, 2000).

Norton, P. (1997). Regional educational technology assistance initiative-phase II: Evaluating a model for statewide professional development. *Journal of Research on Computing in Education, 31*(1), 25–48.

Office of Special Education Programs. (1997). *Report of the office of special education programs: Proceedings of the annual technical assistance and dissemination conference* (7th, Washington, D.C., March 3–5, 1997). (ERIC Document Reproduction Services No. 408 811).

Office of Technology Assessment, U.S. Congress. (1995). *Teachers and technology: Making the connection*. Washington, D.C., Government Printing Office.

Orkwis, R., & McLane, K. (1998). A curriculum every student can use: Design principles for student access. ERIC/OSEP Topical Brief, The ERIC Clearinghouse on Disabilities and Gifted Education, Reston, VA.

Patton, M. Q. (1990). *Qualitative evaluation and research methods* (2nd ed.). Newbury Park, CA: Sage Publications, Inc.

Pugach, M. C. (1991). Uncharted territory: research on the socialization of special education teachers. *Teacher Education and Special Education, 15*.

Raskind, M. H., Goldberg, R. J., Higgins, E. L., & Herman, K. L (1999). Patterns of change and predictors of success in individuals with learning disabilities: Results from a twenty-year longitudinal study. *Learning Disabilities Research & Practice*, *14*(1), 35–49.

Rivera, D. P., & Smith, D. D. (1997). Designing instruction. In: D. P. Rivera & D. D. Smith (Eds), *Teaching Students with Learning and Behavior Problems*. Boston, MA: Allyn and Bacon.

Rosenberg, M. S., Griffin, C. C., Kilgore, K. L., & Carpernter, S. L. (1997). Beginning teachers in special education: A model for providing individualized support. *Teacher Education and Special Education*, *20*(4), 301–321.

Russek, B. E., & Weinberg, S. L. (1993). Mixed methods in a study of implementation of technology-based materials in the elementary classroom. *Evaluation and Program Planning*, *16*(2), 131–142.

Southern Technology Council (STC) (1997). *Making Connections: Seven Principles for State Telecommunications Regulations*. Available: http://www.southern.org/html/stc/pub (1998, February 20).

Szasz, T. S. (1972). *The myth of mental illness*. St. Albans, England: Paladin Press.

U.S. Department of Commerce (2000). Falling through the net: Toward digital inclusion. (http://search.ntia.doc.gov/pdf/fttn00.pdf)

United States Department of Education (1996). *Getting America's schools ready for the 21st century: A report to the nation on technology and education*. Washington, D.C.: U.S. Government Printing Office.

Vanderheiden, Gregg C. (1985). *Computers as augmentative communication systems*.

GENERAL AND SPECIAL EDUCATORS KNOWLEDGE AND PERCEPTIONS OF ASSISTIVE TECHNOLOGIES: ARE WE HEADED IN THE RIGHT DIRECTION?

Jeffrey P. Bakken

ABSTRACT

Students, regardless of disability type can benefit from the implementation of different assistive technologies. The difficulty is that assistive technology devices are as diverse as the needs and characteristics of the children and families who will be using them. Professionals are now responsible for helping children and families select and acquire assistive technology devices and equipment as well as instructing them in their use. In addition, these professionals may lack training in the uses, adaptations, and integration of assistive technology in a variety of activities. This chapter will review past literature that was located addressing assistive technology needs of teachers and the results that have currently been discovered. Next, a study that was implemented with general and special education teachers in regards to their knowledge and perceptions of assistive technology will be discussed. Finally, implications for the future addressing assistive technology will be suggested.

Technological Applications, Volume 15, pages 149–172.

ISBN: 0-7623-0815-x

Individuals with mild disabilities often have difficulty with certain skills like reading, listening, organizing information or writing, and may benefit from the use of assistive technology. Students, regardless of disability type could benefit from the implementation of different assistive technologies.

"Assistive technology" is defined by the Technology-Related Assistance Act of 1988 (Tech Act), P.L. 100-407, and the Individuals with Disabilities Act of 1990, (IDEA), P.L. 101-476, as "any item, piece of equipment, or product system, whether acquired commercially off-the-shelf, modified, or customized, that is used to increase, maintain or improve the functional capabilities of individuals with disabilities." It can describe both devices and services that aid an individual. This is important to understand as many people associate only devices with the term assistive technology and it includes services as well. The definition of assistive technology applies to any technology or product that offers an adult with a disability compensatory techniques. Appropriate assistive technology for students with disabilities can include, but is not limited to, computers, taped books, spellers, tape recorders, readers, calculators and electronic date books. Assistive technologies range from low-tech (simple devices) such as adapted spoons, switch-adapted battery-operated toys, an abacus (for math computation), or a pencil gripper (to help students hold pencils) to high-tech (complex devices) such as augmentative or alternative communication systems, computers, software, peripheral interfaces, sophisticated electronics (i.e. Scan-A-Word reads out loud any typed text, such as magazines, books, newspapers, letters, and forms), and environmental control devices (Lesar, 1998). Equipment or strategies developed for individuals with other disabilities than the child being assessed may also be appropriate. For example, personal FM systems originally developed for the deaf may be an appropriate selection for a child with a learning disability. Technologies to help people with multiple disabilities may indeed help people with learning problems. The key is to determine the functional limitation of the disability and to identify an appropriate accommodation.

Assistive technology, sometimes referred to as adaptive or access technology, includes a whole realm of high and low technology devices designed to increase the independence of individuals with disabilities by enabling them to compensate for deficits, enhance self-confidence, and participate more fully in all settings – work, school, home, and leisure. Assistive technology can enhance the quality of life for a person with a learning disability by enabling the individual to circumvent specific deficits, while capitalizing on given strengths.

Currently, assistive technology devices are as diverse as the needs and characteristics of the children and families who benefit from them (Lesar,

1998). The impetus for using assistive technology evolves from passage of the Technology Related Assistance for Individuals with Disabilities Act of 1988 (P.L. 100-407; commonly referred to as the "Tech Act") which expanded the availability of assistive technology services and devices to all persons with disabilities not just those with low-incidence disabilities (e.g. physical disabilities, deaf and hard of hearing, visually impaired).

Equally important in the process of identifying and selecting appropriate assistive technology is the element of "assistive technology service," defined in the above-referenced legislation as any service that directly assists an individual with a [learning] disability in the selection, acquisition, or use of an assistive technology device. Under the new legislation, assistive technology services to young children includes a number of specific supports. First, a child must have access to an evaluation of his or her technology needs, including a functional evaluation in the child's customary environment. Also indicated in IDEA is the purchasing, leasing, or otherwise providing for the acquisition of assistive technologies. It also includes selecting, designing, fitting, customizing, adapting, applying, maintaining, repairing, or replacing assistive technology devices. In addition, a service consists of coordinating and using therapies, interventions, or services associated with the child's individual education plan (IEP) or programs. Finally, service delivery systems are responsible for providing training and technical assistance to children with disabilities, their families, and the professionals who provide services to them (20 U.S.C. §1401[26]). Professionals are now responsible for helping children and families select and acquire assistive technology devices and equipment as well as instructing them in their use (Lesar, 1998).

However, before the employer or teacher can determine the kinds of assistive technology that will best suit the needs of the student with disabilities, the functional limitations that the individual displays need to be defined. First, what job duties or coursework obligations is the individual expected to perform? In what specific areas is the individual having difficulties? What is it specifically that the individual cannot do or does not do according to the employer's/instructor's expectations? The answers to these questions will determine the kinds of assistive devices that can be put in place to enable the person with mild disabilities to perform the essential functions of the task or meet the requirements of a course.

Lack of knowledge about available devices, funding sources to acquire or lease assistive devices, and evaluating needs to obtain appropriate devices may also inhibit recommendations and actual use of assistive technology (Behrmann, 1995; Blackhurst, 1988; Scherer, 1991). In addition, professionals may not be familiar with the available technologies and lack training in the uses, adaptations, and integration of assistive technology in a variety of activities (Lesar, 1998).

Many professionals directly involved with helping plan and implement assistive technologies have not had any official training in this content area. To do this successfully the teacher must be able to creatively solve problems which is a critically important quality that a special educator should attain (Thompson, Siegel & Kouzoukas 2000). Special educators who are expert problem solvers rely on a variety of tools and strategies, including assistive technologies, in their work (Thompson, Siegel & Kouzoukas 2000). Although assistive technologies are not the answer for every challenge that students with disabilities encounter, special educators who have little knowledge of and/or limited access to them are at risk of becoming ineffective. Without current knowledge of assistive technology, teachers cannot participate meaningfully in solving certain types of problems (Thompson, Siegel & Kouzoukas 2000).

Responding to the times, technology has made considerable advances in helping individuals with disabilities become productive and independent participants in work, classroom, and leisure settings. Recent laws mandating civil rights for those with disabilities can be interpreted to imply that the implementation of technology is a significant opportunity for the provision of equal access. The forces of "equal access," "non-discrimination," and "reasonable accommodations" have created an environment which encourages the use of technology designed to help those with disabilities function on a more equal basis with their non-disabled peers.

The field of special education has a long and rich tradition of using modifications and adaptations to compensate for challenges associated with disabilities (McGregor & Pachuski, 1996). In the last decade, advances in technology have greatly expanded the repertoire of tools and instructional strategies available to enhance the participation of students in a full array of educational settings and activities. However, the benefits of these innovations cannot be realized by students unless teachers are adequately prepared to operate the equipment and integrate it into their classroom routine (McGregor & Pachuski, 1996).

The Americans with Disabilities Act of 1990 (ADA), P.L. 101-336, prohibits discrimination against all individuals with disabilities, requiring both the public and private sector to provide "reasonable accommodations." The application of this mandate is legally interpreted to apply to the acquisition and modification of equipment and devices, such as adaptive hardware and software for computers. The Vocational Rehabilitation Act of 1973, P.L. 93-112, requires that electronic office equipment purchased through federal procurement meet disability access guidelines. The Tech Act's 1994 amendments provide funding to help establish programs to promote the provision of technology-related assistance. The National Literacy Act of 1991, P.L. 102-73, encourages the use of technology in literacy programs.

Independent of the issue of technology, teacher willingness to make adaptations and instructional modifications for students with disabilities has been extensively studied, especially as it relates to mainstreaming and inclusion (e.g. Baker & Zigmond, 1990; Myles & Simpson, 1989; Schumm & Vaughn, 1991; Schumm, Vaughn, Gordon & Rothein, 1994). As a whole, this body of literature suggests that in general education classroom settings, teachers are not always able or willing to modify their instructional strategies or modify the curriculum to accommodate individual students with disabilities. Teachers consistently rate the desirability of making adaptations for individual students higher than the feasibility of actually doing so (Schumm & Vaughn, 1991). In other cases, due to the relative newness of the field of assistive technology, many professionals have not received adequate training to select appropriate technologies and provide effective services. In a national survey of teacher preparation programs in special education, the need for technology training was substantiated by 92% of the respondents (Kinney & Blackhurst, 1987).

Special education settings also have been criticized for their lack of differentiation or individualization in instruction across students (Allington & McGill-Franzen, 1989). Since assistive technology, in many cases, represents a highly specialized knowledge base the extent to which teachers have the necessary skills to enable a student to realize the benefits of his/her equipment in the context of the classroom can be limited without the proper education and training (McGregor & Pachuski, 1996). Available research suggests that many special educators are not particularly well trained or skilled in the applications of assistive technology to the needs of children with disabilities (e.g. Brooks & Kopp, 1989; Parker, Buckley, Truesdell, Riggio, Collins & Boardman, 1990; Todis, 1996).

Thompson, Siegel and Kouzoukas (2000) imply that local school districts, state education systems, and the federal government could certainly do more to provide opportunities for teachers to upgrade their skills with regard to assistive technology. Teacher preparation programs in institutions of higher education must also assume responsibility for the current state of affairs and lack of assistive technology knowledge and expertise of teachers. They also suggest that the teachers themselves must accept some responsibility for their own knowledge development. They suggested that teachers need to make a concerted effort to increase their knowledge and keep up with changes and available strategies in the field.

This chapter will review past literature that was located addressing assistive technology needs of teachers and the results that have currently been discovered. Next, a study that was implemented with general and special education teachers in regards to their knowledge and perceptions of assistive technology will be discussed. Finally, implications for the future addressing assistive technology will be suggested.

Early Childhood Special Education and Assistive Technology

Lesar (1998) distributed and collected the results of surveys from early childhood special education professionals in two southeastern states. The purpose of this study was to collect data and provide information with regard to teacher preparation needs, concerns, and perceived barriers to the use of assistive technology for young children with disabilities. Results indicated that respondents had frequent and significant concerns about their knowledge and utilization of assistive technology. Furthermore, training areas were identified that address the concerns and barriers regarding assistive technology.

Method

Survey packets were mailed to the 169 early childhood special educators. A systematic sampling procedure was implemented to obtain a selected sample of early childhood special education professionals located in two southeastern states. Of the 169 mailed questionnaires, 68 were returned, representing a conditional response rate of 40%. Of the 68 respondents, 97% were Caucasian, 2% were African American, and 2% were from other ethnic backgrounds. On average, respondents had been working as early childhood special education providers for 8.19 years (SD = 5.89) and had been employed in the target preschool program for 5.0 years (SD = 5.22). A 40-item questionnaire focusing on assistive technology preparation, knowledge and usage, training needs, family involvement, and concerns was designed for data collection.

Results

Personal experiences and printed materials were the most frequent types of assistive technology preparational experiences reported by these professionals. Respondents rated their personal experiences as the most highly effective training experience. In contrast, coursework was the least frequent type of assistive technology preparation experience reported by this group. Approximately one-third of the respondents (32%) stated they felt their educational experiences had prepared them to provide and use assistive technology services with young children with disabilities. Conversely, a majority (68%) of these professionals felt unprepared in the use of assistive technology.

These individuals also reported significant deficits in their knowledge of assistive technology. Specific results of knowledge areas include: mobility aids, 64%; communication devices, 59%; agencies that provide services or access to assistive technology, 55%; switch-activated devices, 51%; software, 50%; low-technology devices, 48%; and computers, 47%. A large percentage of the respondents considered themselves to be novices with regard to their knowledge of specific technologies.

Specific problems reported by participants include: (a) training and technical assistance, (b) family involvement, (c) availability and funding of assistive technology, (d) assistive technology assessment processes, and (e) selection and maintenance of assistive technology. Also addressed in this questionnaire were the themes of: (a) concerns about lack of knowledge and access to information on assistive technology devices and services, (b) appropriately matching assistive devices to a specific child, (c) availability of assistive technology devices, and (d) funding sources for obtaining, maintaining, and updating devices. The results of this study suggest that future early childhood special education personnel would benefit from course content and from hands-on experiences that would familiarize them with the knowledge and skills on the uses and benefits of assistive technology.

Teacher Preparedness of Assistive Technology

McGregor and Pachuski (1996) investigated whether teachers in the schools were prepared and supported to implement assistive technologies for their students. This survey was developed because previous investigations regarding teacher readiness to use computers and assistive devices have indicated low levels of knowledge and support (Brooks & Kopp, 1989; Cramer, 1992).

Method

Survey packets were mailed to 600 teachers using assistive technology devices with students throughout Pennsylvania. Of the 600 mailed questionnaires, 366 were returned, representing a conditional response rate of 61%. The respondents were well trained and experienced teachers. Over 70% of the teacher sample had earned a master degree or masters plus additional graduate credits. Years of teaching experience ranged from one to 40 years and respondents taught students representing all levels (preschool through 12th grade).

The survey consisted of 32 items designed to gather information from teachers about their experiences with assistive technology in their classroom. Survey questions were developed after a review of the literature to identify issues that potentially impact teacher use of assistive devices. Questions addressed: (a) professional training and experience of teachers; (b) availability, training, and experience in the use of computers; (c) experience in the use of assistive technology, focusing on the student currently in their classroom who uses assistive technology; (d) teacher perceptions regarding the outcomes of using assistive technology for this student; (e) actions taken by the teacher to help their student utilize assistive technology; and (f) teacher perceptions regarding the supports necessary to use assistive technology.

Results

Students with physical disabilities represented the largest single disability group receiving equipment, followed by students with primary disability classifications of "multiple disability" "mental retardation" and "visual impairment". Over 90% of the students referenced by teachers in their response to this survey used some type of augmentative communication equipment. The most common educational placement for students in this sample was a full time special education classroom located in a public school building attended by both general and special education students ($N = 115$).

Over half of the teachers reported having experience with one to four students who used assistive devices. Almost one-third of the sample had experience with five to twelve different students who used assistive technology. The remaining 10% of the teachers reported that at this point in their career, they had worked with between 12 and 40 different students who used assistive technology. Informal discussions with teachers and a review of the literature yielded a number of barriers that potentially limit the success of technology applications in educational settings (e.g. Blackhurst, 1985; Blackstone, 1990; Campbell, Bricker & Esposito, 1980; Esposito & Campbell, 1987; Parker, Buckle, Truesdale, Riggio, Collins & Boardman, 1990). These will be discussed next.

Over 40% of the respondents reported that time required to become proficient with the equipment presented a substantial barrier to implementation. Similarly, time required to program the equipment and the lack of "loaner" equipment were reported as a barrier by over 30% of the teachers. Finally, over 25% of the teacher respondents reported complexity of use for the student, lack of portability of the equipment, and availability of support to use the equipment as barriers. The most important supports for assistive technology utilization reported by teachers involve time, fiscal support for repair and maintenance, and training.

Implications

These data suggest that general background and proficiency in the use of technology in the classroom for instructional purposes does not minimize the need for specifically focused training on the actual equipment used by a student in the classroom. Assistive technology provides a voice, a means of access, and a tool to compensate for specific challenges associated with their disabilities. Teacher responses in this study suggest a need for interdisciplinary planning teams to develop carefully considered programs and support for the use of assistive devices. It is incumbent, therefore, that planning teams consider the classroom environment, available resources, and competing demands for the classroom teacher's time in developing a support plan that is responsive to an individual student's needs within the context of that particular setting.

Special Educators and Assistive Technology

As part of an assistive technology center planning phase, a survey was implemented to determine: perceived unmet needs of students in regards to assistive technology, perceived individual needs of assistive technology training and competencies, features of an assistive technology center, and how training on assistive technology would be preferred. Thompson, Siegel and Kouzoukas (2000) implemented an assistive technology survey of special educators employed at a university laboratory school, a special education association that serves 17 rural school districts, and a special education association that serves students in a mid-sized city.

Method

Survey packets were mailed to the 234 special educators. Of the 234 mailed questionnaires, 149 were returned, representing a conditional response rate of 64%. Of the 149 respondents, 95% were female and 5% were male. On average, respondents had been working as teachers for 13.49 years (range 0 to 33 years). In addition, teachers reported working with students representing all 13 categories of disabilities under IDEA. A six-page survey was developed through a literature review and solicitation from local administrators of special education programs and university faculty.

Results

None of the teachers predicted that the use of assistive technology would decrease and 95% projected either an increase or great increase in the future. Few teachers believed that assistive technology was an area of expertise for them, a majority perceived that there were critical gaps in their knowledge base, and over one-third reported that they lack basic competence. These data suggest a great need for additional preservice and inservice training of teachers in assistive technology. The perceived need for information is also relevant. A total of 90% of the teachers reported having a moderate or significant need for information on technologies that would assist students with mild disabilities in completing and learning academic tasks.

The teachers also expressed considerable interest in the three areas that crossed disability categories: 86% wanted more information on procedures and tools to assess the assistive technology needs of their students with disabilities; 83% wanted more information on private and public sources to fund assistive technology devices and services; and 78% wanted more information on teaming, specifically, as related to ways that individualized education plan team members could collaborate in addressing the assistive technology needs of students.

The educators did not feel particularly competent in the area of assistive technology and would like for their students to receive a valid assistive technology assessment so that specific needs can be identified and addressed. Many teachers believed that a sizeable number of students could benefit from low cost/low tech assistive technology devices, but did not perceive that there was a great need for high cost/high tech equipment. Most respondents wanted to receive inservice instruction at their school and were receptive to taking courses for graduate credit.

Implications

Overall, it appears that there needs to be an increased effort in teacher training programs as well as within school districts on the area of assistive technology. Not only in the use of assistive technology, but the kinds of assistive technology and ways to assess students for choosing assistive technology.

Up to this point in time, no study was located that surveyed general and special education teachers with regard to their knowledge and perceptions of assistive technology, as well as the ability of their degree granting institution to prepare them for teaching using assistive technology. The purpose of the current investigation was to compare the results of general and special educators in regards to their knowledge and perceptions of assistive technology.

GENERAL AND SPECIAL EDUCATORS KNOWLEDGE AND PERCEPTIONS OF ASSISTIVE TECHNOLOGIES: ARE WE HEADED IN THE RIGHT DIRECTION?

Participants

The participants in this investigation were 100 general and special educators (50 of each) from two Midwestern states. All were graduate students who were taking special education coursework and were currently teaching special education students. A total of 84 were female and 16 were male with a mean age of 31.57 years (SD = 4.52, range of 24–49). The respondents had also been teaching an average of 8.55 years (SD = 4.46, range of 1–26).

Of the 50 general education teachers, 40 were female, and 10 were male with a mean age of 30.64 years (SD = 4.38, range of 24–46). The respondents had also been teaching an average of 7.60 years (SD = 4.25, range of 1–21). Of the 50 special education teachers, 44 were female, and six were male with a mean age of 32.50 years (SD = 4.51, range of 26–49). The respondents had also been teaching an average of 9.50 years (SD = 4.51, range of 3–26).

See Tables 1 and 2 for a comparison.

Procedures

The survey was developed to ascertain what general and special educators in two Midwestern states knew about assistive technology and also their

Table 1. Demographic Data of Entire Sample.

Variable	Total Sample
Gender	84 Female 16 Male
Age	
Mean	31.57
SD	4.52
Range	24–49
Years Teaching	
Mean	8.55
SD	4.46
Range	1–26

Table 2. Demographic Data of Individual Groups.

Variable	General Educators	Special Educators
Gender	40 Female 10 Male	44 Female 6 Male
Age		
Mean	30.64	32.50
SD	4.38	4.51
Range	24–46	26–49
Years Teaching		
Mean	7.60	9.50
SD	4.25	4.51
Range	1–21	3–26

Table 3. Open-ended Questions from Assistive Technology Survey.

Q1 What is your definition of **Assistive Technology?**

Q2 What types of students can benefit from using **Assistive Technology**?

Q3 What are the different types of **Assistive Technologies** you have heard about?

Q4 What are the different types of **Assistive Technologies** you are familiar using?

Q5 What are the different types of **Assistive Technologies** you would feel comfortable teaching a person how to use?

Q6 What is the difference between high-tech and low-tech **Assistive Technology**?

Q7 What are positive outcomes of **Assistive Technology** for students with disabilities?

Q8 What are negative outcomes, if any, of **Assistive Technology** for students with disabilities?

Q9 List the number of **Assistive Technology** courses you have taken: ____

Q10 List the number of courses you have taken that addressed **Assistive Technology**: ____

Q11 List the number of **Assistive Technology** workshops you have attended: ____

Q12 List the number of **Assistive Technology** inservice hours you have received: ____

Table 4. Likert Scale Questions from Assistive Technology Survey.

For the following: 5 = Excellent/Strongly Agree to 1 = Unsatisfactory/Strongly Disagree					
Q13 How knowledgeable do you feel in regards to **Assistive Technology**?	5	4	3	2	1
Q14 How well did your university prepare you with regard to **Assistive Technology**?	5	4	3	2	1
Q15 How important is **Assistive Technology** for the success of students with disabilities?	5	4	3	2	1
Q16 How much support to you receive from your school/district in regards to **Assistive Technology**?	5	4	3	2	1
Q17 How often do you implement any type of **Assistive Technology**?	5	4	3	2	1
Q18 How important is high-tech **Assistive Technology**?	5	4	3	2	1
Q19 How important is low-tech **Assistive Technology**?	5	4	3	2	1
Q20 How much do you value **Assistive Technology** to aid students?	5	4	3	2	1

perceptions about assistive technologies. Teachers in four graduate general and special education classes (two classes from each state) were provided a 20 item questionnaire regarding assistive technology. Twelve questions were open-ended and eight questions required participants to rate items on a 5-point Likert scale. Questions addressed knowledge of assistive technology, kinds of assistive technology, training on assistive technology, and beliefs about assistive technology. See specific questions in Tables 3 and 4.

Results

Results from the survey will be presented next. First, a qualitative analysis of questions 1–8 will be presented. Then a quantitative analysis of questions 9–20 will follow.

Qualitative Analysis of Questions 1–8

Questions 1–8 addressed basic knowledge in regards to assistive technology. Many of the teachers (general and special education) had a difficult time defining assistive technology, however, the definitions from the special education teachers were more specific and none of them responded, "I don't know." Both groups of teachers indicated that all students could probably benefit from the use of assistive technology. The next three questions addressed listing different assistive technologies teachers had heard about, were familiar using, and would feel comfortable teaching someone how to use. Special education teachers listed more types of assistive technologies for all three questions than general education teachers. For both groups of teachers, however, more assistive technologies were listed for having heard about them than for actually being familiar with them or being able to teach someone else how to use them. In addition, special education teachers listed more high-tech devices. In regards to the difference between high-tech and low-tech assistive technologies, both groups of teachers had accurate descriptions with general educators having more inaccurate responses as well as some "I don't know" responses. The last two questions dealt with positive and negative attributes of assistive technology. General and special educators were both able to come up with responses addressing the positive and negative attributes of assistive technology, but the special education teachers were able to come up with more total attributes as well as items that were more specific and addressed a wider range of characteristics. See Tables 5 and 6 for a detailed description of teacher responses.

Quantitative Analysis of Questions 9–20

Questions 9–20 addressed more quantitative aspects in regards to assistive technology. These questions addressed basic knowledge in regards to assistive technology. Each question was answered by circling a value from five (Excellent/Strongly Agree) to one (Unsatisfactory/Strongly Disagree) on a likert scale. For all data a t-test was implemented to check for significant differences between means.

With regard to education on assistive technology 58% of special education teachers, but only 14% of general education teachers indicated they had at least one class on assistive technology. All of the special education teachers had at least one course where assistive technology was addressed while 32% of general education teachers never had even one course where assistive technology was addressed. In regards to workshops attended on assistive technology, all of the special education teachers had attended at least one workshop while 82% of general education teachers indicated they had never attended a workshop on assistive technology. Finally, 70% of special education teachers have never attended an assistive technology inservice while 86% of

Table 5. Results of Open-ended Questions 1–8 for General Educators.

Q1 What is your definition of **Assistive Technology?**

Any mechanical support used to augment and facilitate learning.
Any device that allows a student to be included in regular education.
Anything related to technology.
Any type of high-tech equipment.
I don't know.

Q2 What types of students can benefit from using **Assistive Technology**?

All students.
Special education students.
Students with disabilities.

Q3 What are the different types of **Assistive Technologies** you have heard about?

Highlighters, index cards, graph paper, tape recorders, mini pocket recorders, software programs, laptop computers, electronic notebooks, small word processors, tape recorded lectures or presentations, large print written materials, large print transparencies, material that is scanned, videotape, reduced number of problems per page, basic hand-held calculators, Microsoft Word, communication boards, word processor.

Q4 What are the different types of **Assistive Technologies** you are familiar using?

Highlighters, index cards, graph paper, tape recorders, mini pocket recorders, software programs, laptop computers, electronic notebooks, small word processors, tape recorded lectures or presentations, large print written materials, large print transparencies, material that is scanned, videotape, reduced number of problems per page, basic hand-held calculators, Microsoft Word, communication boards, word processor.

Q5 What are the different types of **Assistive Technologies** you would feel comfortable teaching a person how to use?

Highlighters, index cards, graph paper, tape recorders, mini pocket recorders, software programs, laptop computers, electronic notebooks, small word processors, tape recorded lectures or presentations, large print written materials, large print transparencies, material that is scanned, videotape, reduced number of problems per page, basic hand-held calculators, Microsoft Word, communication boards, word processor.

Q6 What is the difference between high-tech and low-tech **Assistive Technology**?

Computer driven vs. not.
High tech costs more.
Purchased items vs. Home-made things.
Money and ease of use.
Electrical, batteries vs. not mechanical.
I don't know.

Q7 What are positive outcomes of **Assistive Technology** for students with disabilities?

Kids learning and succeeding.
Inclusion in regular education classes.
Children can do more things.
Increased learning opportunities.
Makes learning easier.
Makes homework easier.

Q8 What are negative outcomes, if any, of **Assistive Technology** for students with disabilities?

Amount of training needed to teach and implement devices.
Administrators complaining about cost.
Equipment purchased that does not do what it is supposed to do.

Table 6. Results of Open-ended Questions 1–8 for Special Educators.

Q1 What is your definition of **Assistive Technology?**

Any object that will help students overcome their disability.
Any mechanical support used to augment and facilitate learning.
Any device that helps a child to learn better or more easily.
A device or method used to assist a child to succeed in education or life.
Any device that allows a student to be included in regular education
Any tool used to assist a student in completing a task.

Q2 What types of students can benefit from using **Assistive Technology**?

All Students.
All students, especially those with disabilities.

Q3 What are the different types of **Assistive Technologies** you have heard about?

Highlighters, index cards, frames, color-coding, graph paper, beepers/buzzers, digital clocks, digital watches, talking watches, headphones or earplugs, tape recorders, mini pocket recorders, voice-activated day planners, software programs, pressure-sensitive paper, individual FM amplification devices, laptop computers electronic notebooks, variable speech control tape recorders (VSC), books on disc, CART (Computer-Aided Realtime Translation), tape recorded lectures or presentations, large print written materials, large print transparencies, magnification, enlarged cursor control panels, on-screen keyboards and keyboards that speak, material that is scanned, videotape or videodisc, larger computer monitors, reduced number of problems per page, basic hand-held calculators, hand-held talking calculators, special-feature calculators that enable the user to select speech options to speak, on-screen computer calculator programs with speech synthesis, large screen displays for calculators and adding machines, big number buttons and large keypads, word processor, pencil gripper, read & speak, Microsoft Word, wheel chairs, walkers, braces, magnifying glasses, buttons, switches, Alpha Smart 3000, Dragon Speak, glasses, communication boards, slant boards, touch screen, pictures, levers.

Q4 What are the different types of **Assistive Technologies** you are familiar using?

Highlighters, index cards, color-coding, graph paper, beepers/buzzers, digital clocks, digital watches, talking watches, headphones or earplugs, tape recorders, mini pocket recorders, voice-activated day planners, pressure-sensitive paper, tape recorded lectures or presentations, large print written materials, large print transparencies, on-screen keyboards and keyboards that speak, material that is scanned, videotape or videodisc, larger computer monitors, reduced number of problems per page, basic hand-held calculators, hand-held talking calculators, large screen displays for calculators and adding machines, big number buttons and large keypads, word processor, pencil gripper, Microsoft Word, wheel chairs, braces, magnifying glasses, buttons, switches, Alpha Smart 3000, Dragon Speak, glasses, communication boards, slant boards, touch screen, pictures, levers.

Q5 What are the different types of **Assistive Technologies** you would feel comfortable teaching a person how to use?

Highlighters, index cards, color-coding, graph paper, beepers/buzzers, digital clocks, digital watches, talking watches, headphones or earplugs, tape recorders, mini pocket recorders, pressure-sensitive paper, tape recorded lectures or presentations, large print written materials, large print transparencies, videotape or videodisc, larger computer monitors, reduced number of problems per page, basic hand-held calculators, hand-held talking calculators, big number buttons and large keypads, word processor, pencil gripper, Microsoft Word, wheel chairs, braces, magnifying glasses, buttons, switches, Alpha Smart 3000, Dragon Speak, glasses, communication boards, slant boards, touch screen, pictures, levers.

Q6 What is the difference between high-tech and low-tech **Assistive Technology**?

Technical vs. Non-technical
Computer driven vs. not.
High tech costs more.
Complexity and ease of use.
Money and ease of use.

Table 6. Continued.

Q7 What are positive outcomes of **Assistive Technology** for students with disabilities?
Helping students overcome disabilities.
Helping those with physical disabilities.
Students' quality of life improved.
Kids learning and succeeding.
Inclusion in regular education classes.
Students becoming independent.
Makes learning easier.
Makes homework easier.
Children can do more things.
Increased learning opportunities.
Raises self-esteem.
Q8 What are negative outcomes, if any, of Assistive Technology for students with disabilities?
Becoming too reliant/dependent.
Amount of training needed to teach and implement devices.
Administrators complaining about cost.
Peer responses/reactions.
Equipment purchased that does not do what it is supposed to do.
Lack of proper assessment/matching of technology.

general education teachers have never attended an inservice on assistive technology.

For Question 9, "Assistive technology courses taken?", the mean for general education teachers was 0.14 (SD = 0.351, Range 0–1) while the mean for the special education teachers was 0.72 (SD = 0.701, Range 0–2). Significant differences were found in favor of the number of courses taken by special education teachers $t\ (1, 98) = -5.233$, $p = 0.000$. For Question 10," Courses taken that addressed assistive technology?", the mean for the general education teachers was 0.78 (SD = 0.616, Range 0–2) while the mean for the special education teachers was 2.52 (SD = 1.199, Range 1–5). Significant differences were found in favor of courses taken that addressed assistive technology by special education teachers $t\ (1, 98) = -9.126$, $p = 0.000$. For Question 11 "Assistive technology workshops attended?" the mean for the general education teachers was 0.18 (SD = 0.388, Range 0–2) while the mean for special education teachers was 2.08 (SD = 0.944, Range 1–4). Significant differences were found in favor of assistive technology workshops attended by special education teachers $t\ (1, 98) = -13.161$, $p = 0.000$. For Question 12 "Assistive technology inservice hours received?" the mean for general education teachers was 0.14 (SD = 0.351, Range 0–1) while the mean for special education teachers was 0.30 (SD = 0.463, Range 0–1). Significant differences were being approached for the number of assistive technology inservice hours received by special education teachers t (1, 98) = –1.948, p = 0.054. For Question 13 "How knowledgeable do you feel in regards to Assistive Technology?" the mean for general education teachers was 1.96 (SD = 0.781, Range 1–4) while the mean for special education teachers was 2.64 (SD = 0.921, Range 1–4). Significant differences were found in favor of being knowledgeable in regards to assistive technology by special education teachers t (1, 98) = –3.982, p = 0.000. For Question 14 "How well did your university prepare you in regards to Assistive Technology?" the mean for general education teachers was 1.94 (SD = 0.843, Range 1–4) while the mean for special education teachers was 2.16 (SD = 0.866, Range 1–4). No significant differences were found between means. For Question 15 "How important is Assistive Technology for the success of students with disabilities?" the mean for general education teachers was 4.14 (SD = 0.639, Range 3–5) while the mean for special education teachers was 4.62 (SD = 0.490, Range 4–5). Significant differences were found in favor of the importance of assistive technology for the success of students with disabilities by special education teachers t (1, 98) = –4.213, p = 0.000. For Question 16 "How much support to you receive from your school/district in regards to Assistive Technology?" the mean for general education teachers was 2.18 (SD = 0.800, Range 1–4) while the mean for special education teachers was 2.86 (SD = 0.881, Range 1–5). Significant differences were found in favor of support from the school/district in regards to assistive technology by special education teachers t (1, 98) = –4.040, p = 0.000. For Question 17, "How often do you implement any type of Assistive Technology?" the mean for general education teachers was 3.80 (SD = 0.699, Range 3–5) while the mean for special education teachers was 3.94 (SD = 0.651, Range 3–5). No significant differences were found between means. For Question 18 "How important is high-tech Assistive Technology?" the mean for general education teachers was 3.10 (SD = 0.678, Range 2–4) while the mean for special education teachers was 3.90 (SD = 0.814, Range 2–5). Significant differences were found in favor of the importance of high-tech assistive technology by special education teachers t (1, 98) = –5.339, p = 0.000. For Question 19 "How important is low-tech Assistive Technology?" the mean for general education teachers was 3.56 (SD = 0.733, Range 2–5) while the mean for special education teachers was 4.62 (SD = 0.490, Range 4–5). Significant differences were found in favor of the importance of low-tech assistive technology by special education teachers t (1, 98) = –8.500, p = 0.000. For Question 20 "How much do you value

Table 7. Results of Open-ended Questions 9–12 by Teacher Type.

Variable	General Educators	Special Educators
Q9 Number of **Assistive Technology** courses you have taken:		
Mean	0.1400	0.7200***
SD	0.3505	0.7010
Q10 Number of courses you have taken that addressed **Assistive Technology**:		
Mean	0.7800	2.5200***
SD	0.6158	1.1993
Q11 Number of **Assistive Technology** workshops you have attended:		
Mean	0.1800	2.0800***
SD	0.3881	0.9442
Q12 Number of **Assistive Technology** inservice hours you have received:		
Mean	0.1400	0.3000
SD	0.3505	0.4629

*** Indicates a significant difference between means.

Assistive Technology to aid students?" the mean for general education teachers was 3.80 (SD = 0.783, Range 2–5) while the mean for special education teachers was 4.64 (SD = 0.485, Range 4–5). Significant differences were found in the value of assistive technology to aid students by special education teachers $t\ (1, 98) = -6.453$, $p = 0.000$. See Tables 7 and 8 for data from questions 9–20.

DISCUSSION

The findings of this investigation, although not very generalizeable due to the small sample size, shed some important findings of the state of affairs of assistive technology in regards to knowledge and perceptions of general and special education teachers. It is very apparent that the training of these teachers in assistive technology knowledge and methods has been minimal at best. These data replicate findings that many special educators are not particularly well trained or skilled in the applications of assistive technology to the needs of children with disabilities (e.g. Brooks & Kopp, 1989; Parker, Buckley, Truesdell, Riggio, Collins, & Boardman, 1990; Todis, 1996). The numbers of

Table 8. Results of Likert Scale Questions by Teacher Type.

Variable	General Educators	Special Educators
Q13 Knowledgeable?		
Mean	1.9600	2.6400***
SD	0.7814	0.9205
Q14 Prepared by University?		
Mean	1.9400	2.1600
SD	0.8430	0.8657
Q15 Importance?		
Mean	4.1400	4.6200***
SD	0.6392	0.4903
Q16 Support from School/District?		
Mean	2.1800	2.8600***
SD	0.8003	0.8809
Q17 Often Implemented?		
Mean	3.800	3.9400
SD	0.6999	0.6518
Q18 Important of High-tech?		
Mean	3.100	3.900***
SD	0.6776	0.8144
Q19 Importance of Low-tech?		
Mean	3.5600	4.6200***
SD	0.7329	0.4903
Q20 Value?		
Mean	3.800	4.6400***
SD	0.7825	0.4849

courses taken, courses that addressed assistive technology, and workshops and inservices addressing assistive technology that were attended is alarmingly low for both general and special educators. Without current knowledge of assistive technology, teachers cannot participate meaningfully in solving certain types of problems (Thompson, Siegel, & Kouzoukas 2000). How will that knowledge be obtained if school systems and universities, as well as the teachers themselves, don't get involved?

With regard to knowledge, special education teachers knew more about assistive technologies, could identify more specific types of assistive technologies, were more familiar using and would feel more comfortable teaching someone about specific assistive technologies, and could differentiate high-tech from low-tech assistive technologies with better ease. In addition, special educators were able to identify more positive as well as negative outcomes of assistive technology. The fact that they were able to do this with an average of 0.72 versus 0.14 courses taken, an average of 2.52 versus 0.78 courses taken where assistive technology was discussed, an average of 2.08 versus 0.18 workshops attended and an average of 0.30 versus 0.14 inservices attended signifies that minimal training can have positive affects. However, professionals are now responsible for helping children and families select and acquire assistive technology devices and equipment as well as instructing them in their use (Lesar, 1998). It is very apparent that more training on assistive technology is needed.

Both groups of teachers also indicated that they felt they were not that knowledgeable about assistive technology, not prepared by their degree granting institution or school district on different assistive technologies, yet were implementing them quite often. Yes, teachers need to make a concerted effort to increase their knowledge and keep up with changes and available strategies in the field (Thompson, Siegel, & Kouzoukas 2000), however, the universities and school districts must also be made accountable. Our teacher preparation programs need to incorporate classes on assistive technology and develop classes that teachers in the field can take to further their knowledge bases.

Teachers did report that assistive technology in their opinions was important for the success of students with disabilities, that they valued assistive technologies to aid students and felt that both high-tech and low-tech assistive technologies were important. Special educators did rate the importance of high- and low-tech assistive technologies higher, but that may be do to the fact that these teachers work with these devices more often and are more knowledgeable about exactly what is a high-tech and low-tech assistive technology. The benefits of these innovations, however, cannot be realized by students unless their teachers are adequately prepared to operate the equipment and integrate it into their classroom routine (McGregor, & Pachuski, 1996).

IMPLICATIONS

Although the results of this survey cannot be generalized outside the two areas the populations were from, they do make us think about some important issues concerning assistive technology. It is very apparent that the results from this survey indicate that there has been a lack of information provided to general and special educators in regards to assistive technology. Available research suggests that many special educators are not particularly well trained or skilled in the applications of assistive technology to the needs of children with disabilities (e.g. Brooks & Kopp, 1989; Parker, Buckley, Truesdell, Riggio, Collins, & Boardman, 1990; Todis, 1996). The results from this survey replicate these previous findings.

First, the results indicated that universities/institutions of higher learning are not providing classes or content within other classes in regards to assistive technology. In addition, teachers do not feel knowledgeable or prepared, but see assistive technology as being very important. With the advent of assistive technology plans for all students with special education there is definitely a need for special education teachers to have some expertise in this area. In addition, as more and more students with disabilities are integrated into general education classrooms the general education teachers should have some expertise in this area. As Thompson, Siegel and Kouzoukas (2000) previously discussed, teacher preparation programs must also assume responsibility for the current state of affairs in regards to assistive technology. Starting from the top-level down, universities/institutions of higher learning should provide more assistive technology courses within their teacher education curriculums and more knowledge of assistive technology in their courses.

Second, the results indicated that school districts are not providing workshops or inservices for their teachers in regards to assistive technology. Teachers also reported not getting much support from school districts in regards to assistive technology, yet they implement it quite often. Again, with more integration of special education students and the advent of assistive technology assessments and plans for students with disabilities, teachers (general and special education) need to be current and up-to-date on the most current and effective assistive technologies. School districts need to take the initiative and plan workshops and inservices that specifically address assistive technology to educate its practicing teachers.

Lastly, the results indicated that teachers may not be taking the initiative to learning more about assistive technology. The teachers themselves must accept some responsibility. Teachers need to make a concerted effort to increase their knowledge and keep up with changes and available strategies in the field (Thompson, Siegel, & Kouzoukas 2000). Teachers reported that they valued assistive technology and that high-tech as well as low-tech devices were important, yet they did not feel that knowledgeable and were in fact implementing assistive technologies for students. Others cannot take the full blame, but teachers must also view assistive technology as important and have more self-discipline in taking classes and attending conferences/workshops outside of the school district to further their knowledge base.

Students with exceptional needs deserve the benefit of technology and assistive devices if they will help them to succeed in school, home, and work environments. Assistive technology can make life more convenient for students with disabilities as well as making them more independent. One person cannot do it alone, however, and it is the responsibility of universities, school districts and teachers themselves to make a concerted effort to change and move forward as technology does to meet the demands of students with exceptional needs.

REFERENCES

Allington, R. L., & McGill-Franzen, A. (1989). Different programs, indifferent instruction. In: D. K. Lipsky & A. Gartner (Eds), *Beyond separate education. Quality education for all* (pp. 75–97). Baltimore: Paul H. Brookes.

Baker, J. M., & Zigmond, N. (1990). Are regular education classes equipped to accommodate students with learning disabilities? *Exceptional Children, 54*, 339–348.

Behrmann, M. M. (1995). Assistive technology training. In: K. F. Flippo, K. J. Inge & J. M. Barcus (Eds), *Assistive technology: a resource for school, work, and community* (pp. 211–222). Baltimore: Brookes.

Blackhurst, E. (1985). *Microcomputer technology in special education*. Unpublished manuscript. Lexington, KY: University of Kentucky.

Blackhurst, A. (1988). *Competencies in the University of Kentucky special education microcomputer specialist program*. Paper presented at CEC/TAM Conference on Special Education Technology, Reno, NV.

Blackstone, S. (1990, November). Assistive technology in the classroom: Issues and guidelines. *Augmentative Communication News, 3*(6), 1–4.

Brooks, D. M., & Kopp, T. W. (1989). Technology in teacher education. *Journal of Teacher Education, 40*, 2–8.

Campbell, P. H., Bricker, W., & Esposito, L. (1980). Technology in the education of the severely handicapped. In: B. Wilcox & R. York (Eds), *Quality education for the severely handicapped: the federal involvement* (pp. 223–246). Washington, D.C.: Bureu for the Education of the Handicapped.

Cramer, S. F. (1992). Asssistive technology training for special educators. *Technology and Disability, 1*, 1–5.

Esposito, L., & Campbell, P. H. (1987). Computers and severely and physically handicapped individuals. In: J. D. Lindsey (Ed.), *Computers and exceptional individuals* (pp. 105–124). Columbus, OH: Charles E. Merrill.

Kinney, P., & Blackhurst, A. E. (1987). Technology competencies for teachers of young children with severe handicaps. *Topics in Early Childhood Special Education, 7*(3), 105–115.

Lesar, S. (1998). Use of assistive technology with young children with disabilities: Current status and training needs. *Journal of Early Intervention, 21*(2), 146–159.

McGregor, G., & Pachuski, P. (1996). Assistive technology in schools: Are teachers ready, able, and supported? *Journal of Special Education Technology, 13*(1), 4–15.

Myles, B. S., & Simpson, R. L. (1989). Regular educators' modification preferences for mainstreaming mildly handicapped children. *The Journal of Special Education, 22*, 479–489.

Parker, S., Buckley, W., Truesdell, A., Riggio, M., Collins, M., & Boardman, B. (1990). Barriers to the use of assistive technology with children: A survey. *Journal of Visual Impairment and Blindness*, *84*, 532–533.

Scherer, M. (1991). Assistive technology use, avoidance and abandonment: What we know so far. In: H. Murphy (Ed.), *Proceedings of the sixth annual conference, technology and persons with disabilities* (pp. 815–826). Los Angeles, CA: Office of Disable Student Services, California State University, Northridge.

Schumm, J. S., & Vaughn, S. (1991). Making adaptations for mainstreamed students: General classroom teachers' perspectives. *Remedial and Special Education*, *12*, 18–27.

Schumm, J. S., Vaughn, S., Gordon, J., & Rothlein, L. (1994). General education teachers' beliefs, skills, and practices in planning for mainstreamed students with learning disabilities. *Teacher Education and Special Education*, *17*, 22–37.

Thompson, J. R., Siegel, J., & Kouzoukas, S. (2000). Assistive technology on the eve of the 21st century: Teacher perceptions. *Special Education Technology Practice*, *2*(3), 12–21.

Todis, B. (1996). Tools for the task? Perspectives on assistive technology in educational settings. *Journal of Special Education Technology*, *13*, 49–61.

UNIVERSITY E-MAIL MENTORS FOR ELEMENTARY STUDENTS WITH DISABILITIES: ATTITUDINAL AND LITERACY EFFECTS

Margo A. Mastropieri, Thomas E. Scruggs, Kathy Klingerman, Lisa Mohler, Tara Jeffs, Richard Boon and John Castellani

ABSTRACT

This paper presents the results of an investigation conducted to determine the effects of implementing technology to facilitate written communication for students with learning disabilities and other special needs. The technology was used for several purposes. First, it was used to establish whether a mentor relationship could be established between a university student majoring in education with students with disabilities. Second, it was used to evaluate whether students with disabilities writing would improve because of the additional practice writing e-mail notes back and forth to the university student. Analysis of results suggests that both quantitative and qualitative improvements were made in written communication as a consequence of the e-mail mentor project. Implications for future research and practice are discussed.

Technological Applications, Volume 15, pages 173–184.

ISBN: 0-7623-0815-x

Students with mild disabilities have documented difficulties with literacy in reading and writing, including writing mechanics, fluency, and expression of ideas (Harris & Graham, 1996). Given a history of failure in writing tasks, motivation to write is also frequently low. Initial work using computers to increase writing have yielded preliminary positive findings for increasing length and complexity of final products (e.g. McNaughton, Hughes & Ofiesh, 1997). An untapped area of combining literacy practice with computers is in the systematic use of e-mail between students with disabilities and university mentors. Although several reports of the effects of e-mail communications in educational settings are available, most previous research has focused on normally achieving K-12 students or college undergraduates.

Robinson (1994) described a supported workshop to train 50 K-12 science teachers in the state of Nevada to use e-mail. It was concluded that e-mail can improve science instruction, "principally by giving students greater access to information. It offers the prospect for making education more equitable in areas that do not have access to centers of learning and cultural enrichment" (p. 105). Robinson concluded that as a consequence of the e-mail training project, most of the trained teachers "are on line, sharing information with Nevada schools, schools in other states, and a limited number of schools around the world" (p. 107). No quantitative data were offered in support of the outcomes of the project.

Norton and Sprague (1997) examined the impact of on-line pen pal and collaborative activities related to learning and teaching on three groups of teachers: teachers collaborating as peers, in-service teachers serving as mentors to preservice teachers, and preservice teachers collaborating with in-service teachers. Participants received seven weeks of instruction in use of e-mail accounts, telecommunications, and use of databases that might be integrated into a curriculum. Participants made contact with their partner and collaboratively completed a lesson plan designed by themselves that integrated databases into the curriculum. Participants were expected to complete the on-line assignment in four weeks, sharing rough drafts by uploading and downloading files, using e-mail for asynchronous communication, and using "talk" for synchronous conversations. Final database plans were submitted electronically to the course instructor. A postsurvey was administered during the class period after the lesson plans were due. Results revealed no significant difference in lesson quality created by inservice teams or inservice/preservice teams. Preservice teachers improved in their attitudes toward the possibilities of telecommunications, both with professionals and students, and all participants reported that the collaborative experience was positive. The authors concluded that on-line collaboration between inservice and preservice teachers may have a facilitative effect on preservice supervision.

Pelton and Pelton (1998) described an application of use of the worldwide web, usenets, and e-mail to manage a preservice technology course. Students were taught how to send e-mail messages, how to use the Internet, and were expected to participate in a course newsgroup accessible from the course home page. Students were also asked to create a personal home page that was linked to the course home page. Readings for the course were links from the course readings page to the assigned articles or resources relevant to the current topic. Assignments were also managed via the internet. The authors concluded that student perceptions about their e-mail confidence increased from a mean of 2.65 (on a 5-point scale) at the beginning of the course, to a mean of 3.43 at the end of the course. Similarly, student perceptions of the importance of e-mail went from 3.43 at the beginning of the course to 4.36 on a postsurvey. It was also concluded that the level of faculty involvement in using e-mail positively affects the student's relationship with the professor, and that students rated e-mail use in the top four most useful skills or applications. It was concluded that e-mail increased teacher communication with students.

Wild and Winniford (1993) described a project in which university students in different locations collaborated on a decision making task using e-mail. Students enrolled in a course in quantitative decision-making in Hawaii collaborated with students in the same course in Texas. Students developed a strategy for communication and worked in collaboration with their remote group members to evaluate and solve a complex linear-programming decision-making task. At the end of the project, students made written comments about the remote group collaborative project. Overall the students felt that the project had been very helpful to them, commenting that the project had been "challenging, valuable, and fun" (p. 198). Some challenges involved time constraints, the occasional lack of motivation by a team member, and the fact that the semesters of the two participating universities ended at different times.

Some promising reports of the utility of e-mail communications in educational settings have been found. To date, however, little research is available that describes the utility of e-mail communications in facilitating the literacy skills of students with learning disabilities or other special needs. It was hypothesized that students with mild disabilities would be more motivated to write when the purpose of the writing is to communicate personally using e-mail with a university student mentor. It was further hypothesized that continued personal communication by e-mail would develop the frequency of written communication, and result over time in longer, more fluent, and more expressive written products. This investigation was intended to describe how students with learning disabilities and mild mental retardation responded to an e-mail mentor with respect to: (a) learning how to use computers to access e-mail; (b) composing e-mail messages; (c) changes in length or syntactic complexity of e-mail messages over time; (d) answering relevant school and personal questions; (e) feeling supported by having a university e-mail partner, and (f) feeling more motivated toward computers and e-mail. Seventeen students with mild disabilities (learning disabilities, mild mental retardation, and communication disorders) were matched and randomly assigned to an e-mail first or e-mail second condition. All students were provided with guided and prompted practice in a prescribed sequence of steps, intended to promote use of the computer for composing e-mail messages and for sending messages. All students were pre and post-tested on their attitudes toward the use of computers and e-mail, and all of their e-mail communications were analyzed for substance and style considerations. It was anticipated that students would gain motivation toward both computer usage and school with the developing relationship of the e-mail mentor, as well as improve in composing e-mail messages as a result of the "mentor relationship."

The present investigation, therefore, was intended to determine the impact of establishing an e-mail mentor with a university student on the writing and motivation toward computers and school for elementary aged students with mild disabilities.

METHOD

Participants

Participants included 17 students attending an elementary school in a rural Midwestern area. Fourteen students were characterized as learning disabled, one student as mild mentally retarded, one student as having communication disorders, and one student as having emotional disabilities along with learning disabilities, according to district and state criteria. All students attended the special education resource class and were included in their general education classes for an average of 64.4% of the day. Fourteen students were boys. Students ranged in grade from 3rd to 6th, with IQs ranging from 71 to 105, and reading grade levels ranging from 1.4 to 5.3 in reading decoding and 1.3 to 6.4 in reading comprehension. Students were also ranked by their teacher on "computer use," using the following criteria: (4) student is competent and can independently complete computer assignments and can assist other students with computer-related tasks; (3) student is competent and can complete most computer assignments independently; (2) student is able to complete computer assignments but may require assistance starting or ending a computer task; (1) student requires assistance to complete computer projects. On these criteria, students scored a mean of 2.76 ($SD = 0.92$).

Based on their grade level, achievement data and computer use, students were matched into two comparable groups and then randomly assigned to a "first" or "second" computer use group. Group 1 contained nine students while group 2 contained eight students. Implementation of the e-mail project was staggered, so that the last week of the Group 1 project coincided with the first week of the Group 2 project. In this way, writing samples and post-tests at the end of the project for Group 1 could be compared with writing samples and pretests at the beginning of the project for Group 2.

Materials

Materials for this investigation included three power Macintosh computers that were in the special education resource classroom. Access to the internet was available through the teacher's e-mail account and students used Netscape Navigator as the internet browser and e-mail composer. Two comparable pre- and post-tests were devised for the project. Items were both open ended and Likert-type scale items. Sample items included: How do you feel about going to school? What do you like least about school? How do you feel about using e-mail? How would you feel about spending class time talking to a friend from the university by e-mail on the computer? If you had a friend from the university, what could you learn from that friend?

Procedure

The project took place over a five-month period, including pretesting, training, e-mailing, and a field trip to the university to celebrate students' e-mail usage and meet their mentor. E-mail correspondence between the students with disabilities and the mentor took place over a three-month period. Before the actual e-mail project started all students were pretested and trained by their special education teacher to use the computer and shown how to access e-mail. Students were introduced to the e-mail project by their special education teacher in a whole group lesson for group 1 and group 2 separately. She explained what they would be doing, with whom they would be corresponding, and how the project would work. Questions about e-mail, the Internet, and "Maria" the university student mentor, were answered.

Maria sent the class a letter describing herself and expressing interest in receiving e-mail from the students:

> Hi! My name is Maria. I am a senior in college, which means this is my last year of school. I am going to school to be a teacher. I really like going to school; it is a lot of fun. I have made many friends and learned a lot since I have been at college. During the school months I live in an apartment with three of my friends and our kitten Samantha. My roommates names are Connie, Mary, and Vicky. We have fun living together!
>
> I am from a small town which is about 40 minutes from you. I live on a farm. We grow corn, soybeans, and wheat, and we have two ponies. I have two older brothers named Bill and Michael. Bill is a farmer with my Dad, and Michael is a lawyer. They are both married and live very close to me. I live with my parents and my cat Jerry in a brick house in a wooded area.
>
> I have many hobbies and activities that I like to do. Tennis and volleyball are my favorite sports, and I play them a lot. I also like to ride my bike and go on walks outside. I enjoy reading books, cross-stitching, painting, and spending time with my friends and families. I also play the flute and piano.
>
> I am excited that I get to communicate with each one of you by e-mail. I know we will have lots to talk about and will become great friends! I hope to talk to you soon!
> Maria

Steps for accessing the computer, connecting to the internet, activating the e-mail account, and writing Maria's e-mail address were reviewed. A large poster displayed all of the necessary steps and was placed near the computers, along with a poster displaying the computer rules, and a large chart tracking assignment completion by students. All e-mail steps were modeled and several students had opportunities to practice all of the steps while they were being reviewed for all of the students. One student from each group was appointed as an "e-mail helper" to assist other students in his or her group with computer-related questions during the project as necessary.

Students were provided with special computer times during which they could use the computer to access their e-mail and compose their return messages. Two or three students were scheduled to use the computers simultaneously. Attempts were made to schedule one "computer expert" with a less experienced computer user. During that same time block, the remaining students were scheduled to work directly with the special education teacher or one of the classroom aides.

Each morning the teacher printed out the students e-mail messages from their mentor and placed them on their desks. This was done so students could spend their computer time composing their responses rather than having to spend time reading the message on the screen. The teacher encouraged students to highlight questions that their mentor had asked. Assistance was also provided with spelling and writing but students were not encouraged to change the content or format of their e-mail messages.

RESULTS

Quantitative Results

Number of sentences generated per communication for Group 1 students at the end of their writing project ($M = 9.44$, $SD = 3.7$) was compared with number of sentences generated per communication for Group 2 students at the beginning of their writing project ($M = 5.00$, $SD = 2.6$), and revealed a significant difference, $t(15) = 2.82$ $p = 0.013$. Since these communications were produced during the same week of the semester for both groups, who had been randomly assigned to groups, it appears likely that participation in the writing project facilitated sentence production in Group 1. For both groups together, number of sentences written increased over the course of the project an average of 1.6 sentences ($SD = 3.4$), a difference that approached statistical significance, $t(16) = 1.93$, $p = 0.071$.

Qualitative Results

Student and mentor e-mail communications, student survey written responses, and teacher/mentor anecdotal and evaluative reports were included in qualitative analysis. This analysis revealed an increase in the amount and complexity of student communications, an improvement in the formats and structure of letter writing, and an improvement in attitudes toward writing and computer use.

Over the course of the project, many changes in style and length of the communications were noted. Some of these changes can be observed in the communications of Stacey, who wrote in her first message:

> Hi! I am in 6grade. My name is Stacey. I live in Smithfield. I am thirteen.[1]

Maria replied in a way to encourage more fluency in Stacey's responses:

> Hi Stacey!
>
> Thank you for sending me your message. I was excited to read it! I bet you learn about a lot of neat things in sixth grade. One of my teachers in college is teaching me about bugs. It is a lot of fun! What is your favorite thing to learn about in school?
>
> What are some activities that you like to do? I like to play volleyball and tennis and spend time with my friends and family.
>
> One of my teachers gave me a test this week. I had to study for it a lot, and I was nervous when I took it. Do you ever get nervous when you study for and take a test? I hope to talk to you soon!
>
> Maria

The intention of such responses was to demonstrate an appropriate form to a friendly letter. That is, to communicate interests and activities, and relate these things to the recipient of the communication through questioning. Anticipation of a response was intended to model good communication, and promote a quick response.

Stacey's response was a more elaborate version of the first message, in which she simply reiterated and described more about herself:

> Hi! My name is Stacey. I am in 6grade. My teacher name is Mrs. Chung and my 6grade teacher name is Mr. Bennington. Mrs. Chung is my best friend in school.

However, after further communications by Maria, Stacey's later responses began to take a more mature form:

> Maria,
> I don't like to read. I don't like to do my homework and I have to do homework. I am in choir, band, I go to YES CLUB, I'am the school crossing guard. I hope to talk you soon!
> Stacey

Still later, Stacey maintained her personal responses and began to address the recipient:

> Hi! Maria
> How was weekend. I am happy Sunday I what to chunch and I was happy I got to sing three song. Tomorrow I have to go to band. Bye I have to due my spelling.

Other students, however, wrote more personal messages from the beginning of the project:

> Hi my name is George Kohler. Maria how are you doing. Hay Maria i am doing fine. i have been climbing'running'hopping'joging'playing. what have you been doing Maria? can you tell me please Maria and can you write it on the paper please Maria. I've been eating fruits and vegetables and pizza what's our favorite food's? Maria have you been eating a enough fruit's and vegetables alot of food Maria?

Other changes that were observed over the course of the project involved changes from one sentences per line, to more conventional paragraph form, and use of more complex sentences. For example, in an early letter, William wrote:

> Hi Mare My Name is William
> I like to play sports my favorite sport is soccer.
> I have Dogs thay are black Newfoundland.
> My Dad drives a Semi.
> I have two sisters.

Later communication revealed more complex writing:

> Hi it's William
> I had a basketball game on Saturday. We where down by eleven. We came back and we tide it up. It was tide at 34. I had 18 points. I have a four wheeler. I git to drive the Four Wheeler by my self.

Letters became more personal over time, as students became more comfortable with communicating with Maria, answering her questions and asking their own. One early letter from Scott read:

> I like to play play station. I like reaing. Our Dog is crazy. He barks at cars.

Four weeks later, in response to questions from Maria about teachers, school, and personal interests, Scott wrote:

> Maria,
> Yes I had a teacher scream at me at my last school. It made me want to not talk. I like to play twister. I like to do worksheets. I like to do group projecta with my friends. I like to studey by myself. My favortie thing to do in reading is read. I like to paint in art. I like to play kickball and basketball. I like to sing in music.
> Adam

Another area that was seen to change was sentence complexity. In an early letter, Ralph wrote:

> Hi. My name is Ralph. My bithday is March 6 when is your bithday. What is the color of your eyes my eyes are bule. Do you like college. I know I know I would!
> Sincerely,
> Ralph

Several weeks later, more complex sentences were found in Ralph's communications:

> Dear Maria
> How are you doing? Did I tell you that I like writing to you. It's probably because this is my first time using the internet. I still don't Know what I'm going to do on my B-day. My favorite T. V. show is the simpson's. Try to be fair to your kids, give them the same opportunity. This is something they shuld not do don't let kids throw stuff. My favorite teacher is Mrs. Chung because she takes her time with us. See ya lader alligator, Ralph.

Evaluation of student communications over time was corroborated independently by interviews with the mentor, Maria. Maria said she was at first surprised by the level of writing skills displayed, but she noticed a number of positive changes over the course of the project. Students changed from a one sentence per line format to paragraph format, and began to write longer, more complex sentences by the end. Students changed from writing "general stuff" at the beginning, to more personal information, both concerning themselves as well as Maria. She noticed that letters frequently became more open by the 4th message. Finally, she noticed that messages began to incorporate more of a letter format throughout the course of the project, and speculated that perhaps the special education teacher promoted use of letter formats.

During the first few weeks of the project, the special education teacher communicated that students were excited about being involved in the e-mail project:

> My students are thrilled with this project. They cherish their letters from Maria. This project has also given me the opportunity to address some letter writing skills. We have been printing out their letters and one of my students asked if he could highlight Maria's questions to ensure that he responded to them all. Since then, most of my students have been highlighting their questions from Maria. Occasionally, I have been able to read over a student's shoulder as he or she writes his or her letter and I have been impressed at the care taken at making sure the letter is precise. I have been helping with spelling, if a student asks how to spell a word, but I have not said much about letter content. I made a few general statements to the entire class about paying attention to the questions Maria writes to make sure they are answered. Overall, I am excited about their motivation and desire to write to Maria. Too bad this is not a tool that has been in use all year. It would be interesting to see if their motivation is due to the novelty of this project, interactions with Maria, or the neatness of using e-mail.

The undergraduate mentor reported that the interactive communications with special education students provided her with significant insights into the characteristics of those students that may not have been otherwise attainable. The special education teacher reported at the end of the project:

> The students looked forward to receiving their letters from [their mentor] . . . . The students seemed to be excited about sharing their personal stories, ideas, and feelings . . . . Another positive aspect of the project is that the students were writing three times a week without expressing complaints or negative comments. During other writing experiences, I frequently need to find ways to motivate students to engage with the task. With this project, the students were motivated to write [their mentor] without any prompting from me.

DISCUSSION

Results of this investigation suggest that an interactive e-mail project between students with mild disabilities and university student mentors can improve attitudes and motivation, and enhance writing fluency and expressiveness. One of the significant features of this program is its potential for providing a vehicle for authentic, interactive written communication between students with special needs and university students. Compared with the type of routine classroom writing assignments that students are frequently given, the ongoing e-mail project allowed students to communicate regularly and develop a relationship with someone with whom they would not normally interact. As such, the purposes and rewards of writing could become more apparent to students who are struggling to develop their writing abilities. One helpful feature of e-mail is that communications arrive immediately, and can be responded to immediately. There is no need to wait for letters to arrive in the mail.

At the end of the writing project, students came to the university campus to meet their mentor in person, and had a party celebrating their accomplishments in the project. Students were very happy to meet Maria, and reinforced the teacher's comments that they found the writing project very enjoyable. Along with the students' e-mail comments themselves, these results suggest that interaction with an undergraduate e-mail mentor can be a particularly motivating experience for students with mild disabilities, frequently characterized as having difficulties in written communication, and frequently lacking in motivation and task persistence. Even though few corrections were made on students' independent writing during this project, improvements were seen in fluency, elaboration, and content.

Future research could be helpful in further exploring the benefits of university e-mail mentors. Of particular interest would be the inclusion of larger groups, and the use of standardized measures of writing, to determine the nature of any generalized writing gains. More formal measures of attitude and motivation would also be helpful in documenting the gains of such projects. Additionally, it would be of interest to extend the project to longer implementation periods. This would allow future researchers to test whether the influence is persistent and ongoing, or whether there is a novelty effect, that would ultimately lose influence. In such a case, it would be of interest to know if a single, or multiple interchangeable e-mail mentors are preferable. A single mentor could presumably develop more personal relationships and interactions over a longer time period; while multiple mentors could expand writing skills and promote more generalized responding. Considering the results of the present investigation, it seems that there is a great potential for the use of university students as e-mail mentors to promote written communication skills for students with mild disabilities.

NOTE

1. Names, towns, and other identifying information have been altered in all e-mail communications.

REFERENCES

Harris, K., & Graham, S. (1996). *Making the writing process work: Strategies for composition and self-regulation.* Cambridge, MA: Brookline Books.

McNaughton, D., Hughes, C., & Ofiesh, N. (1997). Proofreading for students with learning disabilities: Integrating computer and strategy use. *Learning Disabilities Research & Practice*, *12*, 16–28.

Norton, P., & Sprague, D. (1997). On-line collaborative lesson planning: An experiment in teacher education. *Journal of Technology and Teacher Education*, *2*, 149–162.

Pelton, L. F., & Pelton, T. W. (1998). Using www, usenets, and e-mail to manage a mathematics per-service course. *Computers in the Schools*, *14*, 79–93.

Robinson, M. (1994). Improving science teaching with e-mail. *Computers in the Schools*, *11*, 94–107.

Wild, R. H., & Winniford, M. (1993). Remote collaboration among students using electronic mail. *Computers in Education*, *21*, 193–203.

PART II:

ISSUES IN LEARNING DISABILITIES

DISCREPANCY MODELS AND THE MEANING OF LEARNING DISABILITY

Kenneth A. Kavale and Steven R. Forness

ABSTRACT

Problems and issues surrounding the use of discrepancy in identifying learning disability are reviewed. Since 1976, discrepancy has been the primary criterion for defining learning disability in practice. In a psychometric and statistical sense, however, issues about the best means for calculating a discrepancy remain unresolved. Another problem involves divergent findings about how systematically and rigorously the discrepancy criterion has been applied in practice. The problems and issues have resulted in questions about the status of learning disability as an independent category of special education. It is possible, however, to demonstrate that learning disability can be reliably differentiated from other conditions and that discrepancy is a major factor in demonstrating the differences. Consequently, it is concluded that discrepancy is a legitimate theoretical concept and should be considered as a necessary criterion for the identification of learning disability.

For some 35 years, the field of learning disabilities (LD) has been witness to enduring debate about the best means for identifying the presence or absence of LD. At the center of the debate is the concept of discrepancy, the difference between expected and actual achievement, which, as a primary (if not sole) criterion, has come under vigorous attack. Since its inception, LD has been defined as originally described in 1968. The substantively unchanged definition

Technological Applications, Volume 15, pages 187–235.
2001 by Elsevier Science Ltd.
ISBN: 0-7623-0815-x

can be found in the Individuals with Disabilities Education Act [IDEA] (Public Law 105-17) and reads as follows:

> The term "specific learning disability" means a disorder in one or more of the basic psychological processes involved in understanding or in using language, spoken or written, which may manifest itself in an imperfect ability to listen, think, speak, read, write, spell or do mathematical calculations. The term includes such conditions as perceptual disabilities, brain injury, minimal brain dysfunction, dyslexia, and developmental aphasia. Such term does not include a learning problem that is primarily the result of visual, hearing or motor disabilities, of mental retardation, of emotional disturbance, or of environmental, cultural, or economic disadvantage (IDEA Amendments of 1997, P.L. 105-17, 11 stat 37 [20 USC 1401(26)]).

What is most curious about this definition is not in what it says but in what it does not say (Kavale & Forness, 2000). The discrepancy concept is not mentioned but yet it has emerged as a primary feature for the identification of LD. The result may be found in two views of LD: a formal, albeit vague, description of LD and a more narrowly focused numerical view of LD. The difficulties lay in the fact that the two views may not be compatible and hence the possibility of misidentification increased.

How did this situation arise? A primary problem is that the federal definition does not stipulate procedural guidelines for LD identification. Consequently, actual practice demands the development of processes to determine who is eligible to receive LD services. The federal definition does not aid the process because it is primarily exclusive in nature by being most descriptive of what LD is not rather than identifying what LD is. The identification process by necessity must consider factors not stipulated in the formal definition. But how was discrepancy selected as the most important identification criterion and how has it become the most controversial? The purpose of this chapter is to answer these questions.

DISCREPANCY CONCEPTS

Intra-individual Differences

In the original Kirk (1962) definition of LD, an important concept articulated was found in the notion of intra-individual differences, the possibility of subaverage functioning in only a few areas with average or above functioning in other areas. Gallagher (1966) termed these "developmental imbalances" that were represented by discrepancies in psychoeducational functioning. One of the first such discrepancies investigated was related to the cognitive abilities of students with LD. Using subtest scores from cognitive assessments like the

Wechsler Intelligence Scale for Children (WISC), patterns of strengths and weaknesses were examined to determine whether the resulting scatter ("profile") differentiated students with LD from other average or low achieving (LA) populations.

Cognitive (Ability) Discrepancy: Profile Analysis

In a comprehensive meta-analytic synthesis, Kavale and Forness (1984) found no WISC profile for students with LD. For example, a discrepancy between Verbal IQ and Performance IQ (V-P split) has been assumed a primary LD characteristic. The difference (PIQ > VIQ) was, on average, only $3\frac{1}{2}$ IQ points that was well below the requisite 11 IQ points necessary for significance. Additionally, although students with LD performed more poorly on Verbal subtests, no Verbal or Performance subtest score fell below the *average* level. Any procedure for calculating WISC inter-subtest variability ("scatter") was not significant and indicated no subtest strength or weakness that distinguished LD performance.

Based on hypotheses about cognitive performance, a number of different subtest score groupings have been proposed to reveal discrepant abilities. One method involves re-categorizing subtest scores exemplified in the proposal by Bannatyne (1968) that included a Spatial, Conceptual and Sequential category. An LD group was presumed to show a Spatial > Conceptual > Sequential pattern, but, while exhibiting the required pattern, the magnitude of the differences were well below the required significance values. A second method seeks a profile that either specifies particular WISC subtest scores as high or low, or identifies particular subtests where students with LD score low. For example, Ackerman, Peters and Dykman (1971) studied the ACID profile (low scores on the Arithmetic, Coding, Information, and Digit Span subtests) but, again, average LD performance did not reach the required level for significant suppression. Similarly, WISC factor scores (e.g. Naglieri, 1981) and WISC patterns (e.g. Myklebust, Bannochie & Killen, 1971) have also been investigated, but in no instance was discrepant performance of students with LD at a level that could be deemed significant.

The long-standing criticism of examining discrepancies appears justified (e.g. Bijou, 1942). In summarizing the available research investigating cognitive performance, Kavale and Forness (1984) concluded that, "Regardless of the manner in which WISC subtests were grouped and regrouped, no re-categorization, profile, pattern, or factor cluster emerged as a clinically significant indicator of LD. In fact, when compared to average levels, the LD group was found to exhibit no significant deviations, and on average, revealed less variability than normal populations" (p. 150).

Ability-Achievement Discrepancy

Origins

The failure to find significant cognitive discrepancies in LD populations along with the desire to reinforce notions about the academic achievement deficits associated with LD directed attention to the possibility of conceptualizing an ability-achievement discrepancy as a primary feature of LD. The ability-achievement discrepancy notion was actually introduced by Bateman (1965) in a definition of LD that included the description of "an educationally significant discrepancy between estimated intellectual potential and actual level of performance related to basic disorders in the learning processes" (p. 220).

The original idea of ability-achievement discrepancy was introduced by Franzen (1920) in the "Accomplishment Quotient" (AQ). The AQ is the ratio of Educational Quotient (EQ) to Intelligence Quotient (IQ). The importance of IQ "lies in its diagnosis of power of adaptation, and it has a high correlation with maximum possible school progress" (p. 434) while the EQ "is the quotient resulting from the division of the age level reached on the test in question by the chronological age of the pupil" (p. 435). The AQ, "the ratio of EQ to IQ gives the percentage of what that child could do, that he has actually done" (p. 436).

In cases where the AQ is less than 90, there exists potential "underachievement." A number of analyses appeared to show that, in general, AQs were typically less than one (e.g. McPhail, 1922; Pintner & Marshall, 1921; Ruch, 1923). The resulting discrepancy was often attributed to lack of effort ("laziness") and "if pupils are pushed to the extreme limit of their ability, the correlation between their educational achievement and their intelligence is not only high but actually reaches unity" (Whipple, 1922, p. 600). There was an associated belief that "bright" students were achieving less than expected, relative to ability, than were "dull" students which was presumed to indicate limited effort from "bright" students. Interestingly, with the 1920s view of intelligence as a fixed entity, IQ was regarded as an index of the upper limit for academic attainment which meant AQs really could not exceed unity ("overachievement"). As suggested by Franzen (1920), "One's differences when EQ is subtracted from IQ are always positive when they are large enough to be significant and small enough to seem spurious when they are negative . . . .

It is safe, therefore, for practical use to assume that the optimum accomplishment is 1.00" (p. 436).

In reality, findings about AQs were suspect because of a number of psychometric and statistical problems. In a comprehensive analysis, Toops and Symonds (1922) discussed a number of flaws with the AQ that were a presage of many later analyses of ability-achievement discrepancy. Many other critiques also appeared that pointed out a number of difficulties. For example, Chapman (1923) demonstrated the unreliability of using difference scores that were based on individual intelligence and achievement test scores. Wilson (1928) suggested that, "Conclusions based on the use of the accomplishment quotient are misleading unless they take into account the reliability of the measures employed, the validity of the measures employed, and the part played by factors determining the intelligence quotient in school achievement under conditions of maximum maturation" (p. 10).

The major statistical criticism of AQ surrounded the operation of the "regression effect" (Crane, 1959; Cureton, 1937). The calculation of AQ assumed a perfect correlation ($r = 1.00$) between IQ and EQ, whereas the real value is closer to 0.60. With this less than perfect correlation between measures, scores above average on one measure will be less superior on the second measure, and at the other end of the continuum, those scores below average on the first measure will be less inferior on the second. Consequently, if AQ does not account for these regression effects, then there will be an over-representation of "bright" students and an under-representation of "dull" students. This situation was demonstrated by Popenoe (1927) who found that, "Instead of each pupil having an equal chance to get a favorable accomplishment quotient, it appears that out of almost five hundred pupils, in no case did an individual having a high intelligence quotient get a favorable accomplishment quotient, and that individuals having a low intelligence quotient obtained accomplishment quotients far above the average. So an AQ of 100 means an entirely different thing in a part of the range from what it does in another" (p. 45). The many difficulties with AQ led to the conclusion that, "the administrative use of the accomplishment quotient is open to serious criticism" (p. 47), and foreshadowed many later issues about the use of ability-achievement discrepancy for LD classification.

Relationship to Learning Disability

The Bateman (1965) notion of discrepancy was never formally incorporated into the federal LD definition. To introduce the discrepancy idea, the then Bureau of Education for the Handicapped issued regulations outlining criteria for the identification of LD. The U.S. Office of Education (USOE; 1976) regulations read as follows:

> A specific learning disability may be found if a child has a severe discrepancy between achievement and intellectual ability in one or more of several areas: oral expression, written expression, listening comprehension or reading comprehension, basic reading skills, mathematics calculation, mathematics reasoning, or spelling. A "severe discrepancy" is defined to exist when achievement in one or more of the areas falls at or below 50% of the child's expected achievement level, when age and previous educational experiences are taken into consideration (p. 52405).

Formula-based Procedures

To assist the identification process, a formula to determine whether a severe discrepancy level (SDL) exists was proposed, but solicited decidedly negative comments and testimonies about its usefulness. For example, Lloyd, Sabatino, Miller and Miller (1977) objected to the use of general intelligence measures and the negative effects of measurement error on accuracy, while Sulzbacher and Kenowitz (1977) objected to the standard 50% discrepancy criterion across academic areas. In an empirical analysis of the SDL, Algozzine, Forgnone, Mercer and Trifiletti (1979) cast doubt on the 50% discrepancy notion "except for children whose measured intelligence falls exactly at 100" (p. 30). Danielson and Bauer (1978) reviewed the issues surrounding formula-based classification procedures and concluded by questioning whether "a technically adequate solution to the problem of LD identification exists" (p. 175).

By 1977, the SDL formula was dropped but not the concept of discrepancy as indicated in regulations stating that:

> A team may determine that a child has a specific learning disability if: (1) The child does not achieve commensurate with his or her age and ability in one or more of the areas listed in paragraph (2) of this section, when provided with learning experiences appropriate for the child's age and ability levels; and (2) The team finds that a child has a severe discrepancy between achievement and intellectual ability in one or more of the following areas: (i) oral expression, (ii) listening comprehension, (iii) written expression, (iv) basic reading skill, (v) reading comprehension, (vi) mathematics calculation, or (vii) mathematics reasoning. (USOE, 1977, p. 6508).

Thus, discrepancy was reinforced as the primary criterion for LD identification (see Chalfant & King, 1976) and, although not given precise specification in a particular formula, became, over time, almost the exclusive factor used for LD determination (Frankenberger & Fronzaglio, 1991; Mercer, Jordan, Allsopp & Mercer, 1996).

QUANTIFYING ABILITY-ACHIEVEMENT DISCREPANCY

With the reinforcement of the idea that a severe discrepancy needs to be demonstrated, individual states were free to choose their own methodology, but the wide variation in procedures chosen introduced a substantial element of arbitrariness to LD identification (Divoky, 1974). Nevertheless, an *in numeris veritas* [in numbers there is truth] mind set developed, and many different means of quantifying a "severe discrepancy" were attempted even though, "there is little reason to believe and much empirical reason to disbelieve the contention that some arbitrarily weighted function of two variables will properly define a construct" (Cronbach & Furby, 1970, p. 79). Thus, the question becomes one of whether two variables (ability and achievement) can be combined to determine the presence or absence of the LD construct.

Grade Level Deviation

The simplest but least sophisticated discrepancy model examines grade level deviations where an expected grade level (EGL) score is compared to an actual grade level (AGL) score and the discrepancy calculated from the EGL–AGL difference. For example, EGL might be based on chronological age (CA), and then discrepancy calculated in terms of "years behind" (EGL = CA − 5). In place of CA, mental age (MA) might be substituted because of its presumed closer relationship between intellectual ability and school achievement (Harris, 1961). The search for increased accuracy led to formulas with additional factors and differential weighting of variables (e.g. Harris, 1971; Monroe, 1932). In any case, when the difference is "significant" (usually 1 to 2 years before grade level), a discrepancy is assumed to exist. The most fundamental problem with this method is the lack of consideration for the level and degree of instruction received.

Expectancy Formulas

Another method of discrepancy calculation involves the use of expectancy formulas that are more comprehensive and include some combination of variables (usually IQ and perhaps CA, MA, years in school [YS], or grade age [GA]). The USOE (1976) formula provides an example:

$$SDL = CA\left(\frac{IQ}{300} + 0.17\right) - 2.5$$

Earlier examples were provided by Bond and Tinker (1973), Harris (1975), and Johnson and Myklebust (1976). The Bond and Tinker formula is:

$$\left(YS \times \frac{IQ}{100} + 1.0 \right) - AGL$$

The Bond and Tucker formula appears to be confounded by the problem of unequal learning rates (Dore-Boyce, Misner & McGuire, 1975) and, unless modified, it is really (along with the Harris formula) a poor predictor that over- and under-identifies students with low and high IQs, respectively (Alspaugh & Burge, 1972; Rodenborn, 1974; Simmons & Shapiro, 1968).

Problems and Issues

The expectancy formula proposed by Johnson and Myklebust (1967) introduced the problem of interpreting ratio scores (AGL/EGL $\times$ 100) where a value less than 90 is considered significant. Because of the absence of an absolute zero and equal intervals, ratio scores possess limited inherent meaning. Only extreme scores are meaningful on what is really an ordinal scale, and a value like 90 cannot be interpreted to mean 90% of average, for example. These difficulties in interpretation were demonstrated by Macy, Baker and Kosinski (1979) where Johnson and Myklebust (1967) discrepancy quotients were found to demonstrate considerable variability across different combinations of age, grade, and academic content areas.

The expectancy formula approach to discrepancy calculation has been roundly criticized (Wilson, Cone, Busch & Allee, 1983). McLeod (1979) discussed the negative influence of measurement errors and the phenomenon of statistical regression. Regression effects may result in an increased probability of misclassification, as pointed out by Thorndike (1963).

> If a simple difference between aptitude and achievement standard scores, or a ratio of achievement to aptitude measure, is completed, the high aptitude group will appear primarily to be "underachievers" and the low aptitude group to be "overachievers". For this reason it is necessary to define "underachievement" as discrepancy of actual achievement from the *predicted* value, predicted upon the basis of the regression equation between aptitude and achievement. A failure to recognize this regression effect has rendered questionable, if not meaningless, much of the research in "underachievement" (p. 13).

When used in actual practice, the application of the expectancy formula approach to discrepancy "yielded strikingly disparate results in terms of the number of children identified as learning disabled by each" (Forness, Sinclair & Guthrie, 1983, p. 111). The prevalence rates ranged from 1% to 37% (Sinclair, Guthrie & Forness, 1984). Confounding this variability was the finding that in a sample of students deemed eligible for LD programs, 64% were not identified by any expectancy formula (Sinclair & Alexson, 1986). Finally, O'Donnell (1980) found that a discrepancy derived from an expectancy formula was not a distinctive characteristic of LD and was equally likely to be found among other students with disabilities.

Score Components

Although discrepancy models were the object of often contentious debate, discrepancy continued to be reinforced as a primary criterion for LD identification primarily because of a desire to reduce the reliance on clinical judgment in LD diagnosis (see Chalfant, 1985). Thus, the continued use of discrepancy in the LD diagnostic process required improved methodology.

The first problem requiring attention was related to the type of test scores included in discrepancy formulas (Hanna, Dyck & Holen, 1979). Age-equivalent scores (e.g. MA) lack a consistent unit of measurement. Even more problematic are grade-equivalent (GE) scores that possess difficulties related to the fact that they ignore the dispersion of scores about the mean and their use creates nonequivalent regression lines between grade and test scores across both grade levels and content areas (Gullicksen, 1950). Consequently, exact values are difficult to achieve and GEs usually involve an excess of extrapolation, especially at upper and lower ends of a scale. The difficulties are compounded because any scores calculated between testing periods (often one year) must be interpolated, but such a calculation is based on the invalid assumption of a constant learning rate. What these problems mean is that individual achievement tests do not exhibit identical GEs (Berk, 1981). The problems associated with GEs may be partially remedied by the use of standard scores that hold the advantage of being scaled to a constant mean (M) and SD but nevertheless still possess limitations and need to be interpreted cautiously (Clarizio & Phillips, 1986).

Standard Score Methods

Standard score (SS) discrepancy models typically involve a direct comparison between intellectual ability and academic achievement (Elliot, 1981; Erickson, 1975; Hanna et al., 1979). In SS discrepancy, scores are first converted to a common metric and then a difference score is calculated. For LD determination, the SSs for ability and achievement most often have an M = 100 and SD = 15 with the discrepancy criterion being a minimum 15 point ability-achievement difference.

Although advancing discrepancy calculation, the SS method is not without limitations. One problem surrounds the assumption that, on average, IQ and achievement scores should be identical (e.g. a student with an IQ of 115 should have a reading or math achievement score of 115). This assumption would be true only if IQ and achievement were perfectly correlated ($r = 1.00$). With the actual correlation about 0.60, the expected achievement score for an IQ of 130 is actually 122, not 130. When dealing with below average IQs, the opposite effect occurs (i.e. an IQ of 85 actually has an expected achievement level of about 88). Thus, the SS approach to discrepancy will always possess a systematic bias (Thorndike, 1963).

Difference Scores

The SS approach produces a difference score that presumably represents an index of discrepancy. The difference score, however, lacks adequate reliability which makes it uncertain whether or not the obtained difference might really have occurred by chance (Feldt, 1967; Payne & Jones, 1957). For example, the satisfactory reliabilities of most IQ and achievement tests (about 0.90) produce a difference score with a reliability of only about 0.75. Measurement error is the primary factor producing the difference score unreliability (see Cronbach, Gleser, Nanda & Rajaratnam, 1972) which may distort the discrepancy calculation as discussed by Thorndike (1963) who concluded that:

> if *nothing* but the errors of measurement in the predictor and criterion were operating, we could *still* expect to get a spread of discrepancy scores represented by a standard deviation of half a grade-unit. We would *still* occasionally get discrepancies between predicted and actual reading level of as much as a grade and a half. This degree of "underachievement" would be possible as a result of nothing more than measurement error (p. 9).

The significantly lower reliabilities of difference scores has been empirically demonstrated (e.g. Algozzine & Ysseldyke, 1981; Reynolds, 1981; Schulte & Borich, 1984) and led Salvia and Clark (1973) to conclude that, "the standard error of measurement for deficit scores is sufficiently large to preclude rigid adherence to deficits as a criterion for learning disabilities" (p. 308).

Regression Methods

With SS models remaining problematic, alternative means of calculating discrepancy were considered. Shepard (1980) suggested a regression discrepancy model to remedy many of the existing problems. The inherent measurement error associated with IQ and achievement tests insures that the phenomenon of statistical regression will occur, especially when dealing with IQ levels outside of a 95–105 range. The regression method involves calculating an equation for IQ and achievement where, "The anticipated [expected] achievement score is the norm for children of the same ability, grade level, and sex" (p. 80). The regression model is the procedure most often recommended for calculating discrepancy (Wilson & Cone, 1984; Reynolds, 1984–1985). Using a variety of statistics, the method takes regression effects into account by calculating a predicted achievement score and a range where the actual achievement score should fall. To illustrate, assume a case where the student's IQ is 112 and the correlation between IQ and achievement is 0.60. With a SS approach (M = 100; SD = 15), the predicted achievement score would be 107.2. Using standard values to denote significant discrepancy level, the number 18.6 is calculated which is then added and subtracted to the predicted achievement score. Finally, assume an actual achievement score of 83 which is below 88.6 (i.e. 107.2 − 18.6) and thus represents a significant (severe) discrepancy.

Evaluating Regression Procedures

Wilson and Cone (1984) argued that "The regression equation approach provides the best method for determining academic discrepancy because unlike other approaches, it considers regression, measurement errors, and evidence" (p. 107). Evans (1990) discussed a number of advantages for the regression approach.

Although the regression discrepancy method provides the best answer to the question, "Is there a severe discrepancy between this child's score on the achievement measure and the average achievement score of all other children with the same IQ as this child?" (Reynolds, 1985, p. 40), fundamental difficulties remain. First, a regression equation requires the choice of a value to denote "severity level" but the vagaries surrounding LD make this choice uncertain. The most usual value chosen is two SDs (gleaned from the historical two SDs below the mean IQ level (i.e. 100) used for the diagnosis of mental retardation [MR]), but while presumably meeting a criterion of "relative infrequency"in the population, the real value remains uncertain because of the lack of a true prevalence rate for LD. This uncertainty may produce classification errors as demonstrated by Shepard (1980) in the form of two types of errors: false-positive (i.e. identifying a student as LD when they are not, in fact, LD) and false-negative (i.e. failing to detect real LD). It was suggested that, "it is likely that the Regression Discrepancy Method falsely labels more normal children as LD than it correctly identifies children who really have a disorder. At the same time, errors of over-identification do not assume that all real instances of LD will be detected" (p. 88).

An Appraisal of Discrepancy Methods

Cone and Wilson (1982) analyzed the four primary models of quantifying a discrepancy and concluded that standard score and regression models are preferred, with regression methods the most preferred because they account for statistical regression effects. This conclusion has been affirmed in other comparative analyses of discrepancy methods (e.g. Bennett & Clarizio, 1988; Braden & Weiss, 1988; Clarizio & Phillips, 1989). Nevertheless, problems remain that make this most defensible method less than perfect with respect to optimal psychometric and statistical considerations (Berk, 1984; Reynolds, 1984–1985). Consequently, when used alone, the inherent difficulties almost make discrepancy "an atheortetical, psychologically uniformed solution to the problem of LD classification" (Willson, 1987, p. 28).

Practical Difficulties

The unresolved technical problems surrounding discrepancy translate into real-world difficulties. Barnett and Macmann (1992) attributed much of the inaccuracy in interpreting discrepancy to basic misunderstandings about test interpretation, statistical significance, confidence intervals, and measurement error. For example, Macmann, Barnett, Lombard, Belton-Kocher and Sharpe (1989) found classification agreement rates about LD or not LD ranging from 0.57 to 0.86 with different discrepancy methods. When both IQ and achievement measures varied in making LD or not LD determination, agreement rates were found to be consistently below 0.25 (Clarizio & Bennett, 1987). Thus, on average, only about 1 in 4 students deemed to possess a "severe" discrepancy would be identified as such when a different set of IQ and achievement tests were used. Macmann and Barnett (1985) affirmed this finding in a computer simulation study that concluded "the identification of a severe discrepancy between predicted and actual achievement was disproportionately related to chance and instrument selection" (p. 371). The consequences become even more problematic in cases where more than one achievement test was administered, and the lowest score among them was selected for use in discrepancy calculation. Sobel and Kelemen (1984) showed how this situation will likely result in a difference between the proportion of students actually classified and the proportion originally expected to be LD. For example, in the case where three achievement tests were administered and the lowest score selected, the original 6.68% LD cases expected to be identified would increase to 12.2%.

Instability of Discrepancy Scores

The potential variability and inaccuracy associated with discrepancy models are made more unsure by findings showing instability in discrepancy scores over time. O'Shea and Valcante (1986) found that patterns of discrepancy between groups of students with LD or LA differed significantly from grade 2 to grade 5. White and Wigle (1986), in a large-scale evaluation of school-identified students with LD, found four different patterns of discrepancy over time. The largest group (40%) revealed no ability-achievement discrepancy at initial placement or at re-evaluation. The next largest groups demonstrated either a pattern of being discrepant at initial placement but not at re-evaluation or, conversely, not being discrepant at initial placement but discrepant at re-evaluation. The smallest group showed a discrepancy at both initial placement and re-evaluation. Such findings suggest that the discrepancy concept is fragile if not applied systematically and rigorously at initial evaluation.

An early survey of 3,000 students with LD in Child Service Demonstration Centers showed that the average discrepancy was only about one year, leading to the conclusion that, "This discrepancy can be interpreted as a moderate retardation, rather than a severe disability" (Kirk & Elkins, 1975, p. 34). In a later similar analysis, Norman and Zigmond (1980) applied the federal SDL formula (1976) and found that, on average, only 47% of students met the criterion. For children aged 6 to 10 years (the likely age range for LD identification), less than 40% met the SDL criterion while the percentage for students aged 15 to 17 was 68%. Although providing greater confidence in the LD status of older students, the smaller percentage of younger children meeting the SDL criterion raises questions about the validity of their initial LD classification.

DISCREPANCY AND THE IDENTIFICATION OF LEARNING DISABILITY

Meeting the Discrepancy Criterion

Shepard and Smith (1983) suggested that, "the validity of LD identification cannot be reduced to simplistic statistical rules" (p. 125), but the inconsistent application of existing criteria creates even more difficulties in the LD diagnostic process. Shepard, Smith and Vojir (1983) found that 26% of already identified students with LD in Colorado revealed no discrepancies while 30% revealed a significant discrepancy with the use of any reading or math achievement test. When the significant discrepancy was validated with a second achievement measure, only 5% of all students with LD possessed a significant discrepancy with two math tests while 27% revealed significant discrepancies with two reading tests. Thus, not only was the discrepancy criterion not supported, but a "below grade level" criterion was not affirmed either, because "Many LD pupils were not achieving below grade level as measured by standardized tests" (p. 317)

Cone, Wilson, Bradley and Reese (1985), in contrast, found that 75% of school-identified students with LD in Iowa met the specified discrepancy criterion. As this LD population continued in school, their achievement levels became increasingly discrepant. In a later analysis, Wilson, Cone, Bradley and Reese (1986) found that the identified students with LD were clearly different from other students with high-incidence mild disabilities in Iowa (e.g. MR and behavior disorders [BD]): "The main factor providing differentiations was discrepancy between achievement and ability" (p. 556). This finding led to the conclusion that students with LD were primarily underachievers defined by the significant discrepancy.

In a later analysis, Valus (1986) found 64% of identified students with LD to be significantly underachieving. In a large-scale analysis of Iowa's LD population, Kavale and Reese (1992) found that 55% met the discrepancy criterion. In different locales, the percentage of students with LD meeting the discrepancy criterion ranged from 32% to 75%. Thus, in any LD population, there will likely be a proportion who do not meet the discrepancy criterion, and these proportions are likely to vary considerably because of varying interpretations found in different locales.

The finding of significant inconsistencies with the percentage of students meeting the discrepancy criterion is common in studies analyzing identified LD populations. For example, McLeskey (1989) found that 64% of an Indiana LD population met the discrepancy criterion. This figure was achieved only after more rigorous and stringent state guidelines for LD identification were implemented, and almost doubled the 33% figure found in an earlier LD sample (McLeskey & Waldron, 1991). In general, about one-third of identified LD samples have been found not to meet the discrepancy criterion (e.g. Bennett & Clarizio, 1988; Dangel & Ensminger, 1988; Furlong, 1988).

Statistical Classification vs. Clinical Judgment

Shepard and Smith (1983) referred to the approximately one-third of already identified students with LD as "clinical cases", meaning that their eligibility was a discretionary judgment made by a multidisciplinary team (MDT) which was at odds with the statistical (i.e. discrepancy) data. This situation may occur because: (a) LD may cause ability level (i.e. IQ) to decline, and if achievement remained at a comparatively low level, then a discrepancy would not exist; (b) intact proficient skills permit the student to "compensate" for the effects of LD which means that achievement test scores may reveal an increase while ability level remained constant; or (c) a "mild" discrepancy was present but not unexplained because of factors such as limited school experience, poor instructional history, behavior problems, or second-language considerations. The essential question becomes: Are such students "truly" LD, or is the inconsistency between MDT decisions and statistical status "truly" misclassification?

The many vagaries associated with "system identification" (Morrison, MacMillan & Kavale, 1985) are the primary reason for the difficulty in decisions about the presence or absence of LD (Frame, Clarizio, Porter & Vinsonhaler, 1982). In analyses of MDT decisions, it appears that LD identification criteria, especially the primary criterion of severe discrepancy, were neither rigorously or systematically applied (Epps, McGue & Ysseldyke, 1982; Furlong & Yanagida, 1985; Furlong & Feldman, 1992). The difficulties begin with the lack of uniformity across educational agencies in setting a "severe" discrepancy criterion (Perlmutter & Parus, 1983; Thurlow & Ysseldyke, 1979). These differences were often then exacerbated by differences in interpreting the existing guidelines (Thurlow, Ysseldyke & Casey, 1984; Valus, 1986). The misapplication of the discrepancy criterion in LD identification was further complicated by external pressures that might include the desire of MDTs to provide special education services, the request of general education teachers to remove difficult-to-teach students, or parental demands for LD placement (e.g. Algozzine & Ysseldyke, 1981; Sabatino & Miller, 1980; Ysseldyke, Christenson, Pianta & Algozzine, 1983).

The many external pressures often become primary considerations because the discrepancy criterion is viewed as too child-centered in a medical model sense and does not permit examination of complex contextual interactions presumed necessary and relevant for valid LD diagnosis (Algozzine & Ysseldyke, 1987). In fact, Gerber and Semmel (1984) argued that an instructional perspective rather than a statistical one should be the basis for determining LD eligibility. They suggested that the teacher become the "test" for determining whether a student has a "real" learning problem. Under such circumstances, it is not surprising to find that MDTs often do not "bother with the data" in the LD identification process (Ysseldyke, Algozzine, Richey & Graden, 1982) and that school personnel were not able to differentiate students with LD based solely on an examination of test scores (Epps, Ysseldyke & McGue, 1984).

Identification and Prevalence

When LD determination is not based on the application of strict criteria, the process may be likened to the U.S. Supreme Court's decision about pornography – "I can't define it, but I know it when I see it." The lack of rigor in the diagnostic process has led to an accelerated rate of LD identification, and LD becoming, by a wide margin, the largest category in special education. Presently, LD accounts for over 50% of all students with disabilities and over 5% of all students in school (U.S. Department of Education, 1999). In commenting on the magnitude of the increase in LD prevalence, MacMillan, Gresham, Siperstein & Bocian, (1996) suggested that, "Were these epidemic-like figures interpreted by the Center for Disease Control, one might reasonably expect to find a quarantine imposed in the public schools of America" (p. 169).

Clearly, fewer students are identified as LD when a strict discrepancy criterion is implemented rigorously (Finlan, 1992), but external factors (e.g. availability of financial resources) may significantly influence the number of students identified as LD (Noel & Fuller, 1985). Forness (1985) showed how special education policy changes in California significantly impacted the number of students identified in the high-incidence mild disability categories. For LD, there was a 156% gain compared to the 104% gain nationally, and when compared to the concomitant losses for MR and BD, led to the conclusion "that California's relatively dramatic increase in children identified as learning disabled may be at the expense of two other related categories" (p. 41). Such disparities among states were not uncommon and led to the conclusion that, "Our results suggest that variation in LD diagnostic levels across states is significantly related to distinctions in diagnostic practice as well as or instead of actual disease prevalence" (Lester & Kelman, 1997, p. 605). In contrast, far greater consistency in classification rates has been found for categories like hearing impairment and physical/multiple disability compared to LD (Singer, Palfrey, Butter & Walker, 1989).

Confounding Among High-Incidence Mild Disabilities

The primary confounding among high-incidence mild disabilities appears to be primarily between LD and MR. For example, MacMillan, et al. (1996) found, among 150 referred students, 43 with IQ levels at 75 or less. Of the 43, only six were classified MR even though they met the requisite IQ cut-off score, while 18 were classified LD primarily because the LD label was viewed as a more acceptable designation. Similarly, Gottlieb, Alter, Gottlieb and Wishner (1994) found that an urban LD sample possessed a mean IQ level that was $1\frac{1}{2}$ SD lower than a suburban comparison sample which led to the conclusion that, "These children today are classified as learning disabled when in fact most are not" (p. 463). This view was affirmed by MacMillan, Gresham and Bocian (1998) who found that out of 61 students classified LD by schools, only 29 met the required discrepancy criterion. They then indicated that, "We did not anticipate the extent to which the process would yield children certified as LD who failed to meet the discrepancy required by the education code" (p. 322). Thus, even though discrepancy remains the primary (and sometimes sole) criterion for LD identification, it seemed to be ignored in actual practice. Gottlieb et al. (1994) suggested "the discrepancy that should be studied most intensively is between the definition of learning disability mandated by regulation and the definition employed on a day-to-day basis in urban schools" (p. 455).

Because "public school practices for diagnosing children with LD bear little resemblance to what is prescribed in federal and state regulations (i.e. administrative definitions) defining LD" (MacMillan et al., 1998, p. 323), the LD population has become increasingly heterogeneous. For example, Gordon, Lewandowski & Keiser (1999) analyzed the problems associated with applying the LD label to "relatively well functioning" students. By failing to rigorously adhere to a discrepancy criterion, students classified as LD may not demonstrate underachievement, a primary LD feature (e.g. Algozzine, Ysseldyke & Shinn, 1982) which then leads to questions about the utility of the LD category (Epps et al., 1984).

DISCREPANCY, LEARNING DISABILITY, AND LOW ACHIEVEMENT

The Minnesota Studies

The vagaries of LD classification, especially the inability to differentiate LD and LA, have been demonstrated in studies conducted by the University of Minnesota Institute for Research on Learning Disabilities ("Minnesota Studies"). Ysseldyke, Algozzine and Epps (1983) analyzed psychometric data obtained from students without LD using 17 operational definitions for LD. In the 248 cases analyzed, 85% met the requirements for one operational definition of LD, while 68% qualified with two or more operational definitions. Only 37% of the sample did not meet any of the 17 operational definitions of LD. A second analysis examined data for students with LD and students with LA to determine how many would qualify with each of the 17 operational definitions of LD used earlier. For the LD group, from 1% to 78% were classified LD with each definition while the LA group was also classified LD from 0% to 71% of the time using each operational definition. Further analysis showed that 4% of the LD group was not classified by any of the 17 operational definitions, while 88% of the LA group qualified as LD using at least one operational definition.

Epps, Ysseldyke and Algozzine (1983), in a similar analysis, examined the number of students identified by each of 14 operational definitions of LD that emphasized discrepancy. In an LD sample, the definitions classified anywhere from 7% to 81% as LD, whereas 5% to 70% of a non-LD group were classified LD using one of these 14 operational definitions. To determine the congruence among the 14 operational definitions, Epps, Ysseldyke and Algozzine (1985) performed a factor analysis and two factors were found. The first factor emphasized LA whereas the second factor was represented by discrepancy. In terms of their respective weights, Factor I accounted for 70% of the variance compared to 16% for Factor II. The difference in explained variance led to the conclusion that LD might be properly conceptualized as a category of LA, rather than one associated with discrepancy.

Epps et al. (1985) also found that knowing how many operational definitions of LD qualified a student provided little assistance in correctly predicting group membership (LD vs. LA). Algozzine and Ysseldyke (1983) also found considerable inaccuracy in decisions about group membership (LD vs. LA), and concluded that, "To make classification dependent on these discrepancies seems somehow arbitrary and capricious" (p. 245). Discrepancy thus appeared to possess limited value, and suggestions about its worth as a criterion for LD identification possessed little merit because "there may be an equally large number of children exhibiting similar degrees of school achievement not commensurate with their measured ability who are not categorized and therefore are *not* receiving special education services even though they are eligible for them under the current conceptual scheme represented by the category of learning disabilities" (p. 246). Thus, the failure to make LD a special education classification defined primarily by the discrepancy criterion suggests that it has not been possible to unequivocally identify a category different from LA, and it therefore might be more appropriate to recognize LA as the major problem.

Learning Disability or Low Achievement

The Minnesota studies appeared to support the view that reliance upon a discrepancy criterion for LD identification may not be defensible because it does not provide a clear distinction between LD and LA. Wilson (1985), however, challenged the idea that the LD category should be eliminated in favor of a more general category like LA. Basically, Wilson suggested that a more generalized classification will do little to eliminate the ambiguities associated with LD. In fact, the Minnesota studies themselves may possess ambiguities and inconsistencies that limited the findings. For example, the Minnesota studies used only a discrepancy criterion for LD identification, and failed to include other components of the federal definition like exclusion which "states that the academic deficit cannot be the result of other possible causes such as emotional and personality factors, cultural deprivation, impaired sensory acuity, or educational deprivation" (p. 45). Since this aspect of the federal definition was not applied, the identification process was necessarily incomplete and restricted.

The other major problem area was related to sampling, specifically the possibility of bias in the Minnesota samples. The final sample used in the Minnesota studies was selected from a much larger population, which raises the question, "Is there evidence to suggest that the selection was random or is there reason to believe that bias may have distorted the findings?" (Wilson, 1985, p. 45). Wilson went on to suggest that, "there is good reason to suspect that selection factors may have produced a disproportionately large number of discrepant achievers in the group of low achievers who were not formally labeled as learning disabled" (p. 46).

In an analysis of a large-scale Iowa sample, Wilson (1985) demonstrated "that the federal definition of learning disabilities can be successfully used, that it can be consistently applied by a large group of special education professionals, that the various components of currently accepted learning disability definitions can provide the basis for discriminating a reasonably unique group of children, and that the exceptions found in this study, and other similar ones, do not automatically invalidate the previous conclusions" (pp. 49–50). The application of both a discrepancy *and* exclusion criterion resulted in a sound foundation for LD classification. As a result, the LD concept appeared defensible because it would be "premature to eliminate it in favor of other concepts that probably have the very same weaknesses" (p. 51). In response, Algozzine (1985) suggested that there was really no "reprieve" for the LD concept. The Wilson statistical analyses were contested, and again it was suggested that LD was a less than viable special education category because "creating the *new* concept of learning disabilities has not reduced the ambiguities, inconsistencies, and inadequacies that existed when low achievement was not a separate diagnostic category" (p. 75).

The Integrity of Learning Disability

The continuing debate about the LD-LA distinction began to erode the integrity of LD. Long-standing critiques of the LD definition (e.g. Reger, 1979; Senf, 1977; Siegel, 1968) evolved into suggestions that LD really does not exist as an independent entity and its depiction as myth (McKnight, 1982), questionable construct (Klatt, 1991), or imaginary disease (Finlan, 1993). The assumption that LD and LA could not be reliably distinguished became part of the conventional wisdom about special education. The primary evidence came from a study by Ysseldyke, Algozzine, Shinn and McGue (1982) showing a substantial degree of overlap between the test scores of LD and LA groups, and a conclusion raising "serious concerns regarding the differential classification of poorly achieving students as either LD or non-LD" (p. 82). Further confirmation was found in a study by Shaywitz, Fletcher, Holahan and Shaywitz (1992) where it was concluded that, "Our findings suggest more similarities than differences between the reading disabled groups" (p. 646). In this study, reading group membership was defined with either a discrepancy criterion (LD) or low achievement (LA) criterion (scoring below 25th percentile in reading). When the LD and LA groups were compared across a number of child-teacher- and parent-based assessments, few differences were found with the major exception being in the ability (i.e. IQ) area. Nearly all the variance between groups was accounted for by IQ, but this may only be a reflection of the way groups were defined.

The findings from these studies have had significant impact and have been reported with remarkable consistency. For example, the Ysseldyke el al. (1982) study has been used to conclude that limited LD-LA differences exist which was exemplified in the following statements gleaned from the literature:

(a) Certain researchers have suggested that LD is largely a category for low-achieving children.
(b) [Ysseldyke et al.] found few psychometric differences between groups of students identified as learning disabled (LD) and low achievers who did not carry the label.
(c) Recent studies of children diagnosed as learning disabled have shown that many such children . . . are virtually indistinguishable from low-achieving non-handicapped peers.

The presumed inability to reliably differentiate LD and LA groups was based primarily on the Ysseldyke et al. (1982) findings showing a large number of identical scores between LD and LA subjects as well as high percentage of overlap between scores. For example, on the *Woodcock-Johnson Psychoeducational Battery*, LD and LA groups were found to have identical scores 33 out of 49 times and an average overlap percentage of 95%. On five other psychoeducational measures, in better than half the cases, there were identical scores between groups and a 96% percentage of overlap. These metrics showing group overlap appeared to be at variance with the reported statistical analyses, however. In a statistical comparison of group performance on Woodcock-Johnson scores, it was found "that on average the LD group performed significantly poorer on 10 of the subtests" (p. 98), while group comparisons of the five other psychoeducational measures showed "that the mean level performance of the LD children was lower on many of the measures, particularly the PIAT, [Peabody Individual Achievement Test], and at times was significantly less than the mean level of their low-achieving peers" (p. 79).

Re-Analysis of the Minnesota Studies

Kavale, Fuchs and Scruggs (1994) re-examined the Minnesota studies using the methods of meta-analysis and demonstrated how the percentage of overlap metric used by Ysseldyke et al. (1982) may have masked real performance differences. The overlap metric used in the Minnesota studies was calculated by using the range of scores found for one group and then comparing how many cases from the second group fell within that same range, but with such a methodology, "The potential bias toward overlap is high because the comparison is based on the variability demonstrated by only one group with the other being forced into that distribution without regard to the characteristics of its own variability" (Kavale et al., 1994, p. 74). The effect size (ES) statistic used in meta-analysis, because it is a standard score (z-score), eliminates potential bias by representing the extent to which groups can be differentiated, or conversely, the degree of group overlap across a dependent variable. For example, an ES of 1.00 indicates that the two compared groups differ by 1 SD at their means and that 84% of one group can be clearly differentiated from the other group with only 16% group overlap.

Using the data from the Ysseldyke et al. (1982) study, Kavale et al. (1994) calculated ESs for 44 comparisons and found an average ES of.338. This means that, on average, it would be possible to reliably differentiate 63% of the LD group. Conversely, 37% could not be differentiated, and this figure represents the degree of overlap which was substantially less than the average 95% reported by Ysseldyke et al. (1982). For the Woodcock-Johnson Cognitive Ability subtests, an average ES of 0.304 was found, while the Achievement subtests provided an average ES of 0.763. With little reason to expect cognitive differences between LD and LA groups, the modest group differentiation was not surprising. On the other hand, almost 8 out of 10 members of the LD group scored at a level that made it possible to discern achievement differences when compared to the LA group. Similar findings emerged with other cognitive and achievement tests. For example, WISC-R comparisons revealed an average ES of.141 (56% level of group differentiation) while PIAT comparisons showed an average ES of 1.14 which indicates that, in almost 9 out of 10 cases (87%), the LD group performance was substantially below that of the LA group. Consequently, "it appears that the lower achievement scores of the LD group are of a magnitude that distinguishes them from their LA counterparts" (Kavale et al., 1994, pp. 74–75).

Algozzine, Ysseldyke and McGue (1995) contested the meta-analytic findings but agreed that students with LD may be viewed as the lowest of the low achievers. They suggested that the difficulty was in interpreting the meaning of that status: "Where we part company is in the inference that because students with LD may be the lowest of a school's low achievers, they necessarily represent a group of people with qualitatively different needs who require qualitatively different instruction" (pp. 143–144). What Algozzine et al. failed to consider, however, were the findings showing minimal group differentiation in the cognitive domain. With essentially no difference in ability but large differences in achievement, the LD group demonstrated significant ability-achievement discrepancies that were not shown by the LA group. Consequently, Kavale (1995) suggested that the LD and LA groups "represent two distinct populations. Because the LD group are lower on achievement dimensions but not on ability, they are, in addition to being the lowest of the low achievers, a different population defined by an ability-achievement distinction represented in a different achievement distribution but not in a different ability distribution" (p. 146).

Re-examining Learning Disability and Low Achievement Samples

In a similar comparison of LD and LA groups that also included comparisons with a MR group defined as IQ < 75, Gresham, MacMillan and Bocian (1996) found an average LD-LA level of differentiation of 61% ($\overline{ES}=0.28$) (compared to 63% reported by Kavale et al. [1994]). The differentiation level for comparisons of LD-MR and LA-MR groups averaged 68.5% and 67.5%, respectively. On achievement measures, LD-LA group comparisons revealed an average ES of 0.39 indicating a 65% level of differentiation, confirming the finding that, "LD children performed more poorly in academic achievement than LA children" (p. 579). The LD group performed most poorly in reading where almost 3 out of 4 students with LD could be reliably differentiated from LA counterparts. The large achievement differences in reading between LD and LA groups was affirmed by Fuchs, Fuchs, Mathes and Lipsey (2000) who found that 72% of the LA group performed better in reading than the LD group ($\overline{ES} = 0.61$). With more rigorous reading assessments, even larger ES differences were found, which "suggest that researchers and school personnel in fact do identify as LD those children who have appreciably more severe reading problems compared to other low-performing students who go unidentified" (p. 95).

Gresham et al. (1996) also investigated cognitive ability (IQ) differences among the three groups. As expected, 94% of the LD group could be reliably differentiated from the MR group. The percentage falls to 73% in differentiating LD and LA groups, suggesting greater overlap in cognitive ability between these two groups. The LA group included by Gresham et al. was, however, defined differently from the earlier Ysseldyke et al. (1982) and Shaywitz et al. (1992) studies: "Our LA group was closer to what might be considered a 'slow learner' group on the basis of their average-level intellectual functioning relative to the LA groups in [the other] studies" (p. 579). Consequently, even though the achievement of the Gresham et al. LA group was depressed, it was not discrepant when compared to their IQ level. In contrast, the Gresham et al. LD group revealed significant ability-achievement discrepancies and were thus properly classified because, "Children with LD perform more poorly in reading than LA children, even when the former group has higher cognitive ability" (p. 580). This finding has been confirmed by Short, Feagans, McKinney and Appelbaum (1986) in an analysis of LD subtypes. In examining reading achievement levels across five groups, it was found that, "the joint application of IQ-and age-discrepancy criteria appeared to be useful for distinguishing between seriously disabled students and those who might be more appropriately classified as slow learners or underachievers" (p. 223). In summary, Gresham et al. (1996) concluded that LD, LA, and MR groups "could be reliably differentiated using measures of cognitive ability and tested academic achievement" (p. 580). When LD was defined with an ability-achievement discrepancy criterion, reliable differences between LD and LA groups emerged, suggesting that discrepancy is an appropriate metric for establishing LD group membership.

LEARNING DISABILITY AND INTELLIGENCE

Is IQ Necessary?

Although empirical evidence appeared to indicate that LD and LA could be reliably differentiated on the basis of a discrepancy criterion, questions about its use continued. One form of such questioning about discrepancy focused attention on IQ and whether it was necessary in definitions of LD. Beginning with findings showing that IQ was not a good predictor when used to define students with reading disability (Siegel, 1988), questions about whether or not IQ was necessary for defining LD arose (Siegel, 1989, 1990). A major problem surrounded IQ tests and what they presumably measure. Stanovich (1991b) concluded that "an IQ test score is not properly interpreted as a measure of a person's potential" (p. 10). Yet, "the LD field has displayed a remarkable propensity to latch onto concepts that are tenuous and controversial .... The LD field seems addicted to living dangerously" (Stanovich, 1989, p. 487). At a practical level, for example, there was controversy about what type of IQ score should be used in discrepancy calculation. Although it was commonly recommended that performance (or nonverbal) IQ be used (e.g. Stanovich, 1986; Thomson, 1982), an equally compelling case could be made for the use of verbal IQ (e.g. Hessler, 1987). Without resolution of these problems, "IQ is a superordinate construct for classifying a child as reading disabled. Without clear conception of the construct of intelligence, the notion of a reading disability, as currently defined, dissolves into incoherence" (Stanovich, 1991a, p. 272).

An ability-achievement discrepancy treats intelligence and achievement as independent variables, but Siegel (1989) suggested that this may not be valid because, "A lower IQ score may be a consequence of the learning disability, and IQ scores may underestimate the real intelligence of the individual with LD" (p. 471). Further confounding was introduced by findings showing that the IQ of students with LD may actually decline over time (Share & Silva, 1987; van den Bos, 1989). If this is a valid finding and also assuming that findings showing students remain close to their original reading levels over time are also valid (Juel, 1988),then discrepancies should increase over time. McLeskey (1992), however, found a negative association between discrepancy level and CA where, "students in the elementary grades were most likely to manifest a severe discrepancy between expected and actual achievement, while high school students were least likely to have such a discrepancy" (p. 18). This finding is in contrast to earlier studies showing greater likelihood of demonstrating a discrepancy in high school (e.g. Norman & Zigmond, 1980).

A partial explanation for these opposing findings may be found in what Stanovich (1986) termed the "Matthew effect" after the Biblical statement (Matthew 13:12) suggesting that each advantage leads to further advantage, or conversely, initial disadvantage multiplies into even more disadvantage. For reading, this means that the poor-get-poorer: "Children with inadequate vocabularies – who read slowly and without enjoyment – read less, and as a result have slower development of vocabulary knowledge, which inhibits further growth in reading ability" (p. 381). Shaywitz et al. (1995), however, found no evidence of a Matthew effect in reading but a modest Matthew effect for IQ in a large-scale LD sample. For both IQ and reading, however, "the influence of the regression-to-the-mean effect tends to mask the relatively small Matthew effect" (p. 902) which suggests that the presumed cumulative disadvantage really refers to the rate of gain or loss in reading ability compared to initial reading level (see Walberg & Tsai, 1983). There are thus complex reciprocal relationships between reading ability and cognitive skills that serve to confound the discrepancy notion because, "the logic of the learning disabilities field has incorrectly assigned all the causal power to IQ. That is, it is reading that is considered discrepant from IQ rather than IQ that is discrepant from reading" (Stanovich, 1991, p. 275).

Dual Criteria Definitions

The problem of confounding between concepts is most likely to arise in situations when constructs are defined on the basis of dual criteria. For example, while the psychometric characteristic IQ has long defined MR (e.g. Hollingworth, 1926), there was a later decision to include a second criterion in the form of adaptive behavior, assessing the effectiveness and degree to which individuals meet standards of self-sufficiency and social responsibility (Heber, 1959). There was, however, concern over the inclusion of adaptive behavior in the MR definition primarily because of measurement issues (Clausen, 1972; MacMillan & Jones, 1972). Specifically, there were no adequate instruments to evaluate adaptive behavior that made it a valid psychometric characteristic comparable to IQ. (Of course, this situation was remedied with instruments like the *AAMD Adaptive Behavior Scale* or the *Vineland Social Maturity Scale*). With only a single reliable and valid measure (i.e. IQ), there would be no means to evaluate *both* criteria and that would create the possibility of students identified as MR who did not meet the dual criteria definition as well as students not identified who would meet the definition were appropriate assessments of *both* criteria available.

When no reliable and valid assessments are available, clinical judgment was substituted but was often unreliable, especially in the "milder" regions of MR. With significant impairment in intellectual ability (IQ < 50), the corresponding adaptive behavior was probably equally impaired and not difficult to judge. As the upper limit of the IQ cut-off was approached (IQ 70–75), however, there was decreasing probability that adaptive behavior would correspond similarly and any clinical judgment would become more problematic.

In defining LD, Kavale and Forness (1985) recommended a dual criteria definition similar to MR that included: (a) significant subaverage academic impairment, and (b) IQ in the average range. The advantage would be that both criteria can be reliably measured and no clinical judgment would be necessary. The two criteria can be readily compared and decision rules adopted to determine when the difference ("discrepancy") was significant. If an additional exclusion criterion was added, then the identification process would avoid the myriad difficulties surrounding attempts to include other definitional parameters (e.g. psychological process deficits, central nervous system dysfunction) that cannot be reliably assessed. For this reason, IQ remains an important component in defining LD.

Defining Learning Disability Without Intelligence

Even though IQ may be considered a necessary variable, Siegel (1989) suggested that the LD field "abandon the use of the IQ test in the definition of learning disabilities . . . the IQ-achievement deviation definition should be abandoned because of its illogical nature" (p. 477). Stanovich (1989) suggested, however, that such a position might be "too extreme" (p. 489) and "perhaps ends up saying too little about too much" (p. 490). Lyon (1989) concluded that, "Siegel has raised some interesting and compelling issues but has confounded her position by taking a narrow conceptual and methodological stance in addressing the relationship between intelligence and the LD definition" (p. 506). Baldwin and Vaughn (1989) suggested that "Siegel's position might be illogical because the reasoning was convoluted and misleading" (p. 513).

Meyen (1989) objected to the suggestion that IQ should be eliminated in the LD definition because, "challenging the use of intelligence measures in defining learning disabilities, in essence, questions the efficacy of the category of learning disabilities itself as a means to identify students who warrant special education services" (p. 482). By eliminating IQ, a situation would be created where, "we would largely serve low achievers and have no basis for determining whether or not a student is achieving at a reasonable level given his or her ability" (p. 482). The result would be an even more contentious debate about LD-LA differentiation. The situation would not be remedied with a different IQ cut-off score which Siegel (1989) suggested as an alternative solution. In applying the discrepancy criterion, there has long been the implicit assumption that the IQ is at an average or above level (i.e. >90) in order to "discriminate between poor achievement that is expected (that is, on the basis of intellectual ability or sensory handicaps) and poor achievement that is not expected (that is, the probable presence of LD)" (Scruggs, 1987, p. 22). With an IQ cut-off score of, for example, 75 (a level closer to the MR criterion) less than average academic achievement would neither be unexpected nor unexplained. There may be a need for special education services, but such a student would not be properly classified as LD.

Confounding With Mental Retardation

The primary difficulty with a lower IQ cut-off score in defining LD would be the potential for even more confounding with MR. The *American Association on Mental Retardation* (Grossman, 1973) shifted the cut-off score for MR from 1 to 2 SD below the mean, that is, an IQ level of 70 instead of 85. Grossman (1983) later suggested that the IQ cut-off might be as high as 75 since IQ score should be viewed as only a rough guideline. Thus, cut-off scores really represent arbitrary statistical decisions rather than levels based on tenets of scientific classification (Zigler & Hodapp, 1986). These arbitrary decisions create a real dilemma surrounding shifting prevalence rates. For example, Reschly (1992) demonstrated that the use of an IQ cut-off of 75 and below results in twice as many individuals potentially eligible as using IQ cut-off of 70 and below. Additionally, more cases fall in the interval 71–75 than in the entire range associated with mild MR (IQ 55–70). For LD with a 75 IQ cut-off score, an additional 22.5% of the population would be eligible (given an "average" IQ level arbitrarily defined at 92.5) with perhaps 3% potentially eligible for either MR or LD. With the discrepancy criterion, eligibility for LD can also be defined with a 1 to 2 SD below the mean level similar to MR (see Mercer et al., 1996). As with MR, however, the choice of criterion level is arbitrary and will also affect prevalence: the smaller the required discrepancy, the larger the prevalence. The current high prevalence rate for LD suggests a tacit decision to include smaller discrepancy criterion levels, but the resulting larger number of students with LD also suggest an increased probability of confounding with MR.

The consequences of the possible confounding between LD and MR are seen in large variations across states in prevalence rates with the typical outcomes showing more LD and less MR than expected (U.S. Department of Education, 1999). Gresham et al. (1996) showed that the percentage of students classified as MR was inversely related to the percentage of students classified as LD ($r = -0.24$). Thus, states serving a small percentage of students with MR classify a larger percentage of students as LD, and vice versa. It is entirely possible then that students with similar cognitive abilities and disabilities are served in one state as LD and another as MR (MacMillan, Siperstein & Gresham, 1996).

The IQ of LD Samples

Although average or above IQ has been considered a prerequisite for LD, there has been a long- standing view that average or above intelligence was not a necessary or desirable criterion (e.g. Ames, 1968; Belmont & Belmont, 1980; Cruickshank, 1977). Support for this view was found in large-scale evaluations of LD populations that have found mean IQ levels in both the low average (IQ 80–90) range (e.g. Koppitz, 1971; Smith, Coleman, Dokecki & Davis, 1977; Wilson & Spangler, 1974) and the lower regions of the average (IQ 90–100) range (e.g. Kirk & Elkins, 1975; McLeskey & Waldron, 1990; Norman & Zigmond, 1980). Additionally, IQ levels for LD samples tended to be quite variable, and anywhere from 10% – 40% of students with LD were found to have IQ scores below 85 (e.g. Gajar, 1980; Kavale & Reese, 1992; Shepard et al., 1983).

To explain why the actual IQ level of students with LD might be below average, Burns (1984) used the bivariate normal distribution to show how LD samples can have average IQ scores well below 100. With the known relationship between IQ and achievement, the average IQ of LD samples will decrease as the correlation between IQ and achievement increases. For example, if cases below a given cut-off (1 SD below the mean) for achievement (e.g. $z = -1.0$) and above a given IQ cut-off (e.g. IQ > 80) are considered, while postulating a correlation of.50 between IQ and achievement, then the average IQ of a sample on the bivariate normal distribution will be about 93. Piotrowski and Siegel (1986), however, suggested that using the bivariate normal distribution to explain mean IQ levels less than 100 for LD samples may not be appropriate. The primary difficulty was found in the use of fixed achievement cut-off scores regardless of IQ score, since achievement is likely to vary as a function of both MA *and* CA. For example, a student with an IQ of 80 and achievement just below the mean (z = –0.05) would meet the discrepancy criterion under the bivariate normal distribution, but in reality, demonstrates almost no underachievement. Conversely, a student with an IQ of 130 and achievement almost 1 SD below the mean (z = –0.95) would, in fact, be underachieving significantly but would not meet the discrepancy criterion for LD. These problems are compounded further as the correlation between IQ and achievement increases. Finally, the bivariate model requires IQ scores to be normally distributed, but this is unlikely given the finding that the IQ of students with LD tend to be less stable (Kaye & Baron, 1987; Lally, Lloyd & Kulberg, 1987).

The Slow Learners

With a proportion of the LD population showing IQ levels falling more than 1 SD below the mean, this group would, at one time, be considered as having borderline MR (see Heber, 1961). As such, this proportion would qualify as "slow learners" (see Ingram, 1935) and likely manifest generalized academic deficiencies. The essential question: Is this group also LD? In some instances the answer might be affirmative, but the majority of this group would probably exhibit academic deficits across *all* achievement domains that would run counter to the assumption that students with LD exhibit achievement deficits in one or more (but not all) academic areas. When all academic achievement areas are equally depressed, the notion of *specificity*, in the sense of intra-individual differences, would not be achieved, even though the idea that LD is associated with a circumscribed set of problems that interfere selectively with academic performance has received support (Stanovich, 1986). Thus, instead of specific LD (as defined in the federal definition), there is a more generalized LD, a concept closer to that described by MR.

The "unexpected" academic failure notion often associated with LD has been the source of other questions about IQ and LD. When identified as LD, a student presumably possesses average or above IQ and meets the discrepancy criterion which then suggests that the cause of the student's academic problems cannot be attributed to low intelligence. On the other hand, the academic deficiencies of students with LA should not be surprising because the achievement problems are consistent with a probable less than average intellectual ability. These differences suggest that the etiology of the two conditions cannot be the same, and, consequently, LD and LA groups possess quantitative and qualitative differences.

LEARNING DISABILITY AND LOW ACHIEVEMENT: QUANTITATIVE OR QUALITATIVE DIFFERENCES?

The Isle of Wight Studies

The origins of assumptions about possible qualitative differences between LD and LA can be found in the Isle of Wight epidemiological studies (Rutter & Yule, 1975; see also Rutter & Yule, 1973; Yule, Rutter, Berger & Thompson, 1974). Essentially, LA (i.e. poor readers) were differentiated into two groups: general reading backwardness (GRB) and specific reading retardation (SRR). The GRB group was defined as reading below expected CA (i.e. no discrepancy between IQ-achievement) while SRR was defined as reading below the grade level predicted from IQ (i.e. the presence of an IQ-achievement discrepancy).

In analyzing the population, Rutter and Yule (1975) found that, while IQ scores were approximately normally distributed, reading achievement scores did not show the same normality because, at the lower end of the distribution, there was an excess of cases ("hump") indicating a greater proportion than the expected 2.3%. This "hump" contained the SRR group whose problems were "specific" to the reading process. As Yule et al. (1974) suggested, "*Extreme* under-achievement in reading occurs at appreciably above the rate expected on the basis of a normal distribution and so constitutes a hump at the lower end of the Gaussian curve . . . . There are no grounds for believing that the hump is anything but a true finding, and the finding implies that there is a group of children with severe and specific reading retardation which is *not* just the lower end of a normal continuum" (p. 10, italics in original).

Rutter and Yule (1975) concluded that, in addition to IQ differences, "Reading retardation is shown to differ significantly from reading backwardness in terms of sex ratio, neurological disorder, pattern of neurodevelopmental deficits and educational prognosis. It is concluded that the concept of specific reading retardation is valid" (p. 195). Rutter (1978) later affirmed the GRB-SRR distinction, and the possibility of etiological differences particularly as manifested in SRR through the minimal brain dysfunction syndrome (Clements, 1966).

Qualitative Distinctions in Mental Retardation

The idea of distributional and etiological differences in a population was first proposed in the MR field. At IQ 50, it becomes possible to distinguish between mild and severe MR. Severe MR (about 25% of the MR population) typically represents "clinical" MR in the sense of possessing, besides limited cognitive ability, CNS pathology and associated clinical indicators. The larger mild MR group typically shows no neurological signs or associated clinical indicators, and represent what is termed "familial" MR (Zigler, 1967). In severe cases, the pathological factors significantly interfere with intellectual development (see Tarjan, Wright, Eyman & Keeran, 1973) to the point that the IQ score distribution becomes distorted as shown by Dingman and Tarjan (1960). In comparing the IQ distributions of low IQ populations (mild and severe) with those of the general population, there was a "hump" at the lower end of the distribution. Above IQ 50, there were few discrepancies between expected and actual percentages derived from the normal distribution, but an excess of cases in the 0–19 IQ and 20–49 IQ ranges. This excess population ("hump") formed its own normal distribution of IQs with a mean IQ of 32 and an SD of 16. Clearly, besides IQ levels, the two groups appeared to differ with respect to both etiology and clinical manifestations (Jastak, 1967; Weir, 1967).

The qualitative differences between the two MR "populations" became the source of debate and evolved into what was termed the "developmental-difference controversy" (Zigler & Balla, 1982). Generally, "this controversy centers around the question of whether the behavior of those retarded persons with no evidence of central nervous system dysfunction is best understood by those principles in developmental psychology that have been found to be generally applicable in explaining the behavior and development of non-retarded persons, or whether it is necessary to involve specific differences over and above a generally lower rate and asymptote of cognitive development" (p. 3).

Qualitative Distinctions in Learning Disability

Because of the developmental-difference controversy, the related GRB-SRR distinction also became contentious. For example, many studies have failed to find a GRB-SRR bimodal distribution (e.g. Rodgers, 1983; Share, McGee, McKenzie, Williams & Silva, 1987; Stevenson, 1988). Van der Wissel and Zegers (1985) suggested that no "hump" was found in the distribution because it may, in reality, be an artifact resulting from floor and ceiling effects associated with the reading measures used. Using designs where students differed in reading level but were comparable in age (CA design) or comparable in reading level but varying in age (reading-level match design), a number of studies failed to demonstrate that SRR groups (achievement scores below levels predicted by IQ, i.e. discrepant) could be differentiated from a GRB group (depressed achievement not discrepant from IQ) (Fletcher, Espy, Francis, Davidson, Rourke & Shaywitz, 1989; Fletcher, Francis, Rourke,

Shaywitz & Shaywitz, 1992; Foorman, Francis, Fletcher & Lynn, 1996; Rispens, van Yperen & van Duijn, 1991; Share & Silva, 1986; Siegel, 1992; Shaywitz et al., 1992; Vellutino, Scanlon & Lyon, 2000). Consequently, IQ was not a major factor associated with SRR which was interpreted to mean that SRR was not a discrete entity, but rather:

> It occurs along a continuum that blends imperceptibly with normal reading ability. These results indicate that no distinct cut-off exists to distinguish children with dyslexia clearly from children with normal reading ability; rather, the dyslexic children simply represent a lower portion of a continuum of reading capabilities (Shaywitz et al., 1992, p. 148).

Rutter (1990) suggested that, "the crucial test of the SRR hypothesis, however, does not depend on the presence or absence of a hump in the distribution but whether the correlates and outcomes of SRR serve to differentiate the syndrome from [GRB]" (p. 637). A number of studies have failed to differentiate GRB and SRR groups along a number of dimensions, however. For example, GRB groups (i.e. no IQ – achievement discrepancy) performed no differently on independent measures of reading achievement or on assessments of the cognitive abilities presumed to underlie the ability to learn to read (Fletcher et al., 1994; Francis et al., 1996; Morris et al., 1998; Share, McGee & Silva, 1989; Stanovich & Siegel, 1994). With respect to gender differences, the presumption of a disproportionately larger number of boys than girls in SRR groups has not received support (Pennington, Gilger, Olson & DeFries, 1992; Share et al., 1987; Shaywitz, Shaywitz, Fletcher & Escobar, 1990). Finally, SRR groups were presumed to have a poorer educational prognosis than GRB groups (Yule, 1973), but little evidence supports the validity of this assumption (Francis et al., 1996; Shaywitz et al., 1992; Vellutino et al., 1996; Share et al., 1989).

In a summary of the available evidence, Fletcher et al. (1998) concluded that:

> Under no circumstances is wholesale use of IQ test for learning disabilities justified. We have shown numerous problems with the discrepancy model, regardless of whether IQ tests or some other measures are used to operationalize the aptitude index. It is not the use of the IQ test that creates the problems with discrepancy. Classifications of children as discrepant vs. low achievement lack discriminative validity (p. 200).

It was then suggested that the discrepancy criterion not be part of the LD identification process primarily because "it is not the score on the IQ test that identifies the child as having learning disabilities, but rather the score on the test of academic achievement that identifies the child with LD" (p. 201). Similarly, Aaron (1997) concluded that "a review of research in the area of reading disabilities indicates that classifying poor readers on the basis of a discrepancy formula into LD and non-LD categories lacks validity on both theoretical and empirical grounds" (p. 488). As an alternative, Aaron suggested a more pragmatic approach based on the Reading Component Model that identifies the source of the reading problem for all students and then focuses remedial efforts on that particular source.

STATUS OF DISCREPANCY IN THE IDENTIFICATION OF LEARNING DISABILITY

Confounding of LD & RD

The discrepancy criterion for LD identification has thus been seriously challenged with some anticipating its "impending demise" (Aaron, 1997). One difficulty, however, is in interpreting what it actually means for the LD construct. The many analyses investigating the discrepancy criterion focused attention on the GRB-SRR distinction where, in both cases, the primary problem was an inability to read. Consequently, there was little question about the presence of reading disability (RD), but the presence or absence of LD was not really considered except by implication. Although students with LD are quite likely to manifest reading difficulties, they may not, and this makes any generalization from a GRB-SRR comparison suspect. The primary difficulty is conceptual and relates to the fact that if RD and LD are considered equivalent constructs, then the scientific law of parsimony (Occam's Razor) is violated (Swanson, 1991). There appears, however, to be a decided tendency to view LD and RD as the same entity as evidenced in statements like, "It is time to abandon the discrepancy-based classification of poor readers into LD and non-LD categories and expand the boundaries of LD to include all children who experience difficulties in learning to read" (Aaron, 1997, p. 488). Such a suggestion, however, would result in even greater confounding between concepts.

The same possible confounding may be found with RD itself. The focus on GRB and SRR as discrete groups tends to obscure the fact almost all students with SRR could be classified as GRB, while half of students with GRB can be classified as SRR (Hinshaw, 1992). In considering SRR itself, there are questions about its relationship to *dyslexia*, a RD equally difficult to define (Benton & Pearl, 1978). The many similarities between these conditions raise the question as to whether SRR and dyslexia are the same thing (Yule & Rutter, 1976). Regardless of how this question may be answered, any response including discussion about LD seems inappropriate since it represents a different (and distinct) phenomenon that may or may not include students with any type of reading problem.

Thus, both LD and RD are complex entities and eliminating the discrepancy criterion does not appear to be a sensible solution for resolving these complexities. Any suggested alternative criterion, as, for example, the Reading Component Model proposed by Aaron (1997) does not appear to be a viable solution in any significant sense unless also associated with a belief that LD is not a legitimate construct. When the LD concept is not considered legitimate, the general theme of any discussion usually calls for stopping unnecessary and unjustified LD labeling, and focus instead on the difficulties of some students in learning to read by providing them with what are believed to be effective and responsive interventions (e.g. McGill-Franzen & Allington, 1993; Christensen, 1992; Swerling & Sternberg, 1996). As suggested by Aaron (1997), "When the discrepancy formula disappears from the educational scene, so will the concept of LD. After 40 years of wandering in the wilderness of learning disabilities, we are beginning to get a glimpse of the promised land" (p. 489). Whether or not the disappearance of the discrepancy formula leads to a "promised land" is certainly moot and by no means the sole solution to vexing problems surrounding LD.

Operational Definitions of LD

A major roadblock to resolving problems surrounding LD is the lack of a precise description of LD (Kavale & Forness, 1995). Yet, the LD field has witnessed unprecedented growth and has accomplished this expansion by using not an agreed upon formal definition but rather a number of operational definitions that stipulated rules about how the term LD was to apply in a particular case if specified operations or actions yielded certain characteristic results. Thus, a construct like LD has a set of operations that define it and knowing these operations is to presumably understand the concept fully (Benjamin, 1955).

For LD, the primary operation has been the application of a discrepancy criterion. Beginning with the USOE (1976) regulations and reaffirmed in several proposed operational definitions (e.g. Chalfant & King, 1976; Shaw, Cullen, McGuire & Brinckerhoff, 1995), discrepancy has emerged as the primary (and sometimes sole) criterion used in the LD identification process. The LD identification process, however, may be more difficult and complicated than it appears when using operational definitions. For example, a first problem surrounds the theoretical validity of operations. In a scientific sense, an operational definition must bear a logical and rational relationship to the verified theoretical constructs stipulated in a formal definition (Bergmann, 1961). For LD, a problem is created because the federal definition makes no mention of discrepancy (Kavale, 1993). Consequently, this lack of correspondence between definitions means that essentially two distinct views of LD are being presented: a formal representation described in the federal definition and an operational representation.

Operational Definitions: Meaningfulness and Significance

The lack of correspondence is a consequential problem. The result is an increased probability that the operational definition may not be justified and lead to potentially meaningless and insignificant operations (Deese, 1972). Although it is possible to operationalize most phenomenon, the resulting operational definition may not be valid. The operations specified may not actually "define" anything but merely state procedures required to test for the presence of the phenomenon to which the operations refer (Kavale, Forness & Lorsbach, 1991). For example, assume an operational definition of LD that is based on the hypothetical *Learning Disability Coefficient* (LDC) whose procedures require a calculation including an individual's white blood cell count multiplied by body weight in ounces, divided by head circumference in centimeters. Although possible to calculate, the LDC would possess little meaning or significance because the available information about LD clearly indicates that the LDC does not "fit" with any validated knowledge.

A less obvious example surrounds the different meanings that may be conveyed when different operational indicators are chosen. For example, discrepancy is typically defined as the difference between ability and achievement, but any number of ability (i.e. IQ) measures and probably even a greater number of achievement measures might be chosen for comparison. The problem is that when different combinations of measures are used to define discrepancy, it is not at all evident that the assessments are operationally, and thus, definitionally equivalent (Deese, 1972). When combined with the demonstrated psychometric and statistical difficulties, the conceptual problems may produce difficulties in "making sense" of the resulting discrepancy.

The use of operational definitions is thus neither a simple nor straightforward process but one that requires significant theoretical verification. Unfortunately, the LD field has not achieved the necessary level of verification primarily because discrepancy was so quickly embraced as a practical solution: "The debate that rages over what LD might be and the lack of consensus over the importance of any given variable is in sharp contrast to the relative unanimity regarding discrepancy. The consensus regarding discrepancy as the primary identification variable for LD has entrenched discrepancy to the point where it now represents the foundation concept for LD diagnosis" (Kavale &

Forness, 1994, p. 23). In fact, discrepancy has become a deified concept as evidenced in its ascension to the status of "imperial criterion" (Mather & Healey, 1990) and a reified concept as seen in its elevation to an almost tangible property of students with LD (Kavale, 1987). Such deification and reification does not appear justified given the fact discrepancy itself is a hypothetical construct defined by hypothetical constructs (see Messick, 1981) resulting in the possibility that, in a theoretical sense, discrepancy may be a "fictitious concept" (Hempel, 1952).

Relationship Between Discrepancy and LD

The deification and reification of discrepancy has obscured some fundamental considerations. One such consideration surrounds the relationship between discrepancy and LD. With discrepancy the primary (and sometimes sole) criterion used for LD identification, there has been an implicit assumption that discrepancy represents the operational definition of LD. In reality, "Discrepancy is best associated with the concept of underachievement. This is true now and has historically been the case" (Kavale, 1987, p. 18). In a theoretical context, Shepard (1983, 1989) argued that discrepancy is the operational definition of underachievement. Thus, when a student meets the discrepancy criterion, what is being affirmed is underachievement, not LD. The scientific law of parsimony would again suggest that underachievement and LD are not the same thing. To avoid confounding, when the discrepancy criterion is met, the proper conclusion is that underachievement has been identified. If it is believed that underachievement is associated with LD (certainly a valid assumption), then discrepancy becomes a necessary but not sufficient criterion for LD identification (Kavale, 1987; Reynolds, 1984–1985).

Within the context of LD identification, discrepancy and the documentation of underachievement should represent only the initial step in diagnosis (Kavale & Forness, 1994). Discrepancy is fundamental to the identification process because the basis for later LD determination is established. Although discrepancy possesses a number of well-documented psychometric and statistical problems, they have been satisfactorily addressed, and a technically defensible procedure to indicate the presence or absence of underachievement has been identified. The findings from large-scale investigations appear to have affirmed the relationship between discrepancy and underachievement, and the ability to reliably differentiate LD (i.e. students who meet the discrepancy criterion) from LA (i.e. students who do not meet the discrepancy criterion). Although critical as the initial step in LD determination, discrepancy need not be elevated to deified or reified status but rather viewed simply as the most useful means for defining underachievement that represents a sound foundation for defining LD.

When discrepancy is placed in proper perspective along with an understanding that LD was never meant to be solely or primarily underachievement, attention needs to be directed at what else should be considered in the identification process in order to capture the complex and multivariate nature of LD (Kavale & Nye, 1991). Kavale and Forness (1995) suggested a way the process might proceed. The initial step is the formulation of foundation principles that is basically a conceptual activity directed at developing a theoretical framework for elucidating the basic nature of LD. From these foundation principles, it would be possible to develop a comprehensive view of LD.

Discrepancy in Context

Kavale and Forness (2000) elucidated the process further by proposing an operational definition in the form of a hierarchical scheme where each level depicts a decision point in the determination of LD. The scheme includes five levels where the first consists of documenting an ability – achievement discrepancy to determine the presence or absence of underachievement. The next four levels focus on additional stipulated criteria (e.g. psychological process deficits, exclusion), and a final LD designation is predicated on a student successfully proceeding through each level. The process ceases if a student cannot meet the requisite criterion at any level. With its initial position, the ability – achievement discrepancy provides the foundation and would be further strengthened if the difference score was based on the most reliable *total* IQ score and *total* achievement test score. In this way, a too narrowly focused discrepancy, as in, for example, a comparison between a Performance IQ and a Social Studies achievement subtest, would be eliminated, and questions about whether or not such a discrepancy can be properly termed underachievement avoided. The outcome would be a more comprehensive view of LD and greater confidence in declaring that a student is "truly" LD.

CONCLUSION

Discrepancy is an important and legitimate concept associated with LD. Beginning with its status as a measure of educational progress, discrepancy evolved into an index of underachievement. When underachievement is considered an integral component of the LD construct, discrepancy, its operational definition, becomes a factor in the identification process. Although open to debate about statistical and psychometric properties, these issues may be satisfactorily resolved and discrepancy calculation made adequate and appropriate for use in LD identification. Because of its efficiency and convenience, discrepancy has become the primary criterion in the LD identification process, but this status has led to a number of difficulties, most noticeably the confounding between LD and LA. In fact, there is little confounding and discrepancy remains a useful component for LD identification. The critical point is that discrepancy not be used as the sole criterion in LD identification. As the operational indicator of underachievement, discrepancy documents the presence of underachievement, not LD. Because LD and underachievement are not equivalent, the task becomes one of deciding what other factors need to be considered in the identification process to be confident about the presence or absence of LD. When placed in a larger, appropriate context, arguments against the use of discrepancy in LD identification diminish. It would, therefore, be an error to eliminate discrepancy as a factor in LD determination. The task is to use discrepancy in a manner where it is not LD itself but rather only part of a more comprehensive identification process.

REFERENCES

Aaron, P. G. (1997). The impending demise of the discrepancy formula. *Review of Educational Research, 67*, 461–502.

Ackerman, P., Peters, J., & Dykman, P. (1971). Children with learning disabilities: WISC profiles. *Journal of Learning Disabilities, 4*, 150–166.

Algozzine, B. (1985). Low achiever differentiation: Where's the beef? *Exceptional Children, 52*, 72–75.

Algozzine, B., Forgnone, C., Mercer, C., & Trifiletti, J. (1979). Toward defining discrepancies for specific learning disabilities: An analysis and alternatives. *Learning Disability Quarterly, 2*, 25–31.

Algozzine, B., & Ysseldyke, J. E. (1981). An analysis of difference score reliabilities on three measures with a sample of low-achieving youngsters. *Psychology in the Schools, 18*, 133–138.

Algozzine, B., & Ysseldyke, J. E. (1981). Special education services for normal children: Better safe than sorry. *Exceptional Children, 48*, 238–243.

Algozzine, B., & Ysseldyke, J. (1983). Learning disabilities as a subset of school failure: The over-sophistication of a concept. *Exceptional Children, 50*, 242–246.

Algozzine, B., & Ysseldyke, J. E. (1987). Questioning discrepancies: Retaking the first step 20 years later. *Learning Disability Quarterly, 10*, 301–312.

Algozzine, B., Ysseldyke, J. E., & McGue, M. (1995). Differentiating low-achieving students: Thoughts on setting the record straight. *Learning Disabilities Research and Practice, 10*, 140–144.

Algozzine, B., Ysseldyke, J. E., & Shinn, M. R. (1982). Identifying children with learning disabilities: When is a discrepancy severe? *Journal of School Psychology, 20*, 299–305.

Alspaugh, J., & Burge, P. (1972). Determination of reading expectancy. *The Journal of Experimental Education, 40*, 1–5.

Ames, L. B. (1968). A low intelligence quotient often not recognized as the chief cause of many learning difficulties. *Journal of Learning Disabilities, 1*, 735–738.

Baldwin, R. S., & Vaughn, S. (1989). Why Siegel's arguments are irrelevant to the definition of learning disabilities. *Journal of Learning Disabilities, 22*, 513, 520.

Bannatyne, A. (1968). Diagnosing learning disabilities and writing remedial prescriptions. *Journal of Learning Disabilities, 1*, 242–249.

Barnett, D. W., & Macmann, G. M. (1992). Aptitude-achievement discrepancy scores: Accuracy in analysis misdirected. *School Psychology Review, 21*, 494–508.

Bateman, B. D. (1965). An educational view of a diagnostic approach to learning disabilities. In: J. Hellmuth (Ed.), *Learning Disorders* (Vol. 1, pp. 219–239). Seattle, WA: Special Child Publications.

Belmont, I., & Belmont, L. (1980). Is the slow learner in the classroom learning disabled? *Journal of Learning Disabilities, 13*, 496–499.

Benjamin, A. C. (1955). *Operationism.* Springfield, IL: C. C. Thomas.

Bennett, D., & Clarizio, H. (1988). A comparison of methods for calculating a severe discrepancy. *Journal of School Psychology, 20*, 359–369.

Benton, A. L., & Pearl, D. (Eds). (1978). *Dyslexia: An appraisal of current knowledge*. New York: Oxford University Press.

Bergmann, G. (1961). Sense and nonsense in operationism. In: P. G. Frank (Ed.), *The Validation of Scientific Theories* (pp. 45–56). New York: Collier.

Berk, R. A. (1981). What's wrong with using grade-equivalent scores to identify learning disabled children? *Academic Therapy.*

Berk, R. A. (1984). An evaluation of procedures for computing an ability-achievement discrepancy score. *Journal of Learning Disabilities, 17*, 262–266.

Bijou, S. W. (1942). The psychometric pattern approach as an aid to clinical analysis – a review. *American Journal of Mental Deficiency, 46*, 354–362.

Bond, G. L., & Tinker, M. (1973). *Reading difficulties: Their diagnosis and correction* (3rd ed.). New York: Appleton-Century-Crofts.

Braden, J. P., & Weiss, L. (1988). Effects of simple difference versus regression discrepancy methods: An empirical study. *Journal of School Psychology, 26*, 133–142.

Burns, E. (1984). The bivariate normal distribution and the I.Q. of learning disability samples. *Journal of Learning Disabilities, 17*, 294–295.

Chalfant, J. C., & King, F. L. (1976). An approach to operationalizing the definition of learning disabilities. *Journal of Learning Disabilities, 9*, 228–243.

Chalfant, J. C., & King, F. S. (1976). An approach to operationalizing the definition of learning disabilities. *Journal of Learning Disabilities, 9*, 228–243.

Chapman, J. C. (1923). The unreliability of the difference between intelligence and educational ratings. *Journal of Educational Psychology, 14*, 103–108.

Christensen, C. A. (1992). Discrepancy definitions of reading disability: Has the quest led us astray? A response to Stanovich. *Reading Research Quarterly, 27*, 276–278.

Clarizio, H. F., & Bennett, D. E. (1987). Diagnostic utility of the K-ABC and WISC-R/PIAT in determining severe discrepancy. *Psychology in the Schools, 24*, 309–315.

Clarizio, H. F., & Phillips, S. E. (1986). The use of standard scores in diagnosing learning disabilities: A critique. *Psychology in the Schools, 23*, 380–387.

Clarizio, H. F., & Phillips, S. E. (1989). Defining severe discrepancy in the diagnosis of learning disabilities: A comparison of methods. *Journal of School Psychology, 27*, 383–391.

Clausen, J. (1972). Quo vadis, AAMD? *Journal of Special Education*, *6*, 51–60.

Clements, S. D. (1966). *Minimal brain dysfunction in children: Terminology and identification*. (NINDS Monograph No. 3, U.S. Public Health Service Publication No. 1415). Washington, D.C.: Department of Health, Education, and Welfare.

Cone, T. E., & Wilson, L. R. (1981). Quantifying a severe discrepancy: A critical analysis. *Learning Disability Quarterly*, *4*, 359–371.

Cone, T. E., Wilson, L. R., Bradley, C. M., & Reese, J. H. (1985). Characteristics of LD students in Iowa: An empirical investigation. *Learning Disability Quarterly*, *8*, 211–220.

Crane, A. R. (1959). An historical critical account of the accomplishment quotient idea. *British Journal of Educational Psychology*, *29*, 252–259.

Cronbach, L. J., & Furby, L. (1970). How we should measure "change" – or should we? *Psychological Bulletin*, *74*, 68–80.

Cronbach, L. J., Gleser, G. C., Nanda, H., & Rajaratnam, N. (1972). *The dependability of behavioral measurements: Theory of generalizability for scores and profiles*. New York: Wiley.

Cruickshank, W. M. (1977). Myths and realities in learning disabilities. *Journal of Learning Disabilities*, *10*, 51–58.

Cureton, E. E. (1937). The accomplishment quotient technique. *Journal of Experimental Education*, *5*, 315–326.

Dangel, H. L., & Ensminger, E. E. (1988). The use of discrepancy formulas with LD students. *Learning Disabilities Focus*, *4*, 24–31.

Danielson, L. C., & Bauer, J. N. (1978). A formula-based classification of learning disabled children: An examination of the issues. *Journal of Learning Disabilities*, *11*, 163–176.

Deese, J. (1972). *Psychology as science and art*. New York: Harcourt Brace Jovanovich.

Dingman, H. F., & Tarjan, G. (1960). Mental retardation and the normal distribution curve. *American Journal of Mental Deficiency*, *64*, 991–994.

Divoky, D. (1974). Education's latest victim: The "LD" kid. *Learning*, *3*, 4–8.

Dore-Boyce, K., Misner, M., & McGuire, D. (1975). Comparing reading expectancy formulas. *The Reading Teacher*, *29*, 8–14.

Elliot, M. (1981). Quantitative evaluation procedures for learning disabilities. *Journal of Learning Disabilities*, *14*, 84–87.

Epps, S., McGue, M., & Ysseldyke, J. E. (1982). Inter-judge agreement in classifying students as learning disabled. *Psychology in the Schools*, *19*, 209–220.

Epps, S., Ysseldyke, J. E., & Algozzine, B. (1983). Impact of different definitions of learning disabilities on the number of students identified. *Journal of Psychoeducational Assessment*, *1*, 341–352.

Epps, S., Ysseldyke, J. E., & Algozzine, B. (1985). An analysis of the conceptual framework underlying definitions of learning disabilities. *Journal of School Psychology*, *23*, 133–144.

Epps, S., Ysseldyke, J. E., & McGue, M. (1984). "I know one when I see one" – Differentiating LD and non-LD students. *Learning Disability Quarterly*, *7*, 89–101.

Erickson, M. T. (1975). The z-score discrepancy method for identifying reading-disabled children. *Journal of Learning Disabilities*, *8*, 308–312.

Evans, L. D. (1990). A conceptual overview of the regression discrepancy model for evaluating severe discrepancy between I.Q. and achievement scores. *Journal of Learning Disabilities*, *23*, 406–412.

Feldt, L. A. (1967). Reliability of differences between scores. *American Educational Research Journal*, *4*, 139–145.

Finlan, T. G. (1992). Do state methods of quantifying a severe discrepancy result in fewer students with learning disabilities? *Learning Disability Quarterly, 15*, 129–134.

Finlan, T. G. (1994). *Learning disability: The imaginary disease*. Westport, CT: Bergin & Garvey.

Fletcher, J. M., Francis, D. J., Rourke, B. P., Sheywitz, S. E., & Shaywitz, B. A. (1992). The validity of discrepancy-based definitions of reading disabilities. *Journal of Learning Disabilities, 25*, 555–561, 573.

Fletcher, J. M., Francis, D. J., Shaywitz, S. E., Lyon, G. R., Foorman, B. R., Stuebing, K. K., & Shaywitz, B. A. (1998). Intelligent testing and the discrepancy model for children with learning disabilities. *Learning Disabilities Research and Practice, 13*, 186–203.

Fletcher, J. M., Shaywitz, S. E., Shankweiler, D. P., Katz, L., Liberman, I. Y., Steubing, K. K., Francis, D. J., Fowler, A. E., & Shaywitz, B. A. (1994). Cognitive profiles of reading disability: Comparisons of discrepancy and low achievement definitions. *Journal of Educational Psychology, 86*, 6–23.

Foorman, B. R., Francis, D. J., Fletcher, J. M., & Lynn, A. (1996). Relation of phonological and orthographic processing to early reading: Comparing two approaches to regression-based, reading-level- match design. *Journal of Educational Psychology, 88*, 639–652.

Forness, S. R. (1985). Effects of public policy at the state level: California's impact on MR, LD, and ED categories. *Remedial and Special Education, 6*, 36–43.

Forness, S. R., Sinclair, E., & Guthrie, D. (1983). Learning disability discrepancy formulas: Their use in actual practice. *Learning Disability Quarterly, 6*, 107–114.

Frame, R. E., Clarizio, H. F., Porter, A. F., & Vinsonhaler, J. R. (1982). Inter-clinician agreement and bias in school psychologists' diagnostic and treatment recommendations for a learning disabled child. *Psychology in the Schools, 19*, 319–327.

Francis, D. J., Shaywitz, S. E., Stuebing, K. K., Shaywitz, B. A., & Fletcher, J. M. (1996). Developmental lag versus deficit models of reading disability: A longitudinal, individual growth curves analysis. *Journal of Educational Psychology, 88*, 3–17.

Frankenberger, W., & Fronzaglio, K. (1991). A review of states' criteria and procedures for identifying children with learning disabilities. *Journal of Learning Disabilities, 24*, 495–500.

Franzen, R. (1920). The accomplishment quotient: A school mark in terms of individual capacity. *Teachers College Record, 21*, 432–440.

Fuchs, D., Fuchs, L. S., Mathes, P. G., & Lipsey, M. W. (2000). Reading differences between low-achieving students with and without learning disabilities: A meta-analysis. In: R. Gersten, E. P. Schiller & S. Vaughn (Eds), *Contemporary Special Education Research: Syntheses of the Knowledge Base on Critical Instructional Issues* (pp. 81–104). Mahwah, NJ: Lawrence Erlbaum.

Furlong, M. J. (1988). An examination of an implementation of simple difference score distribution model in learning disability identification. *Psychology in the Schools, 25*, 132–145.

Furlong, M. J., & Feldman, M. G. (1992). Can ability-achievement regression to the mean account for MDT discretionary decisions? *Psychology in the Schools, 29*, 205–212.

Furlong, M. J., & Yanagida, E. H. (1985). Psychometric factors affecting multidisciplinary team identification of learning disabled children. *Learning Disability Quarterly, 8*, 37–46.

Gajar, A. H. (1980). Characteristics across exceptional categories: EMR, LD, and ED. *Journal of Special Education, 14*, 165–173.

Gallagher, J. J. (1966). Children with developmental imbalances: A psychoeducational definition. In: W. M. Cruickshank (Ed.), *The Teacher of Brain-injured Children* (pp. 20–34). Syracuse, NY: Syracuse University Press.

Gerber, M. M., & Semmel, M. I. (1984). Teacher as imperfect test: Reconceptualizing the referral process. *Educational Psychologist, 19*, 137–148.

Gordon, M., Lewandowski, L., & Keiser, S. (1999). The LD label for relatively well- functioning students: A critical analysis. *Journal of Learning Disabilities, 32*, 485–490.

Gottlieb, J., Alter, M., Gottlieb, B. M., & Wishner, J. (1994). Special education in urban America: It's not justifiable for many. *Journal of Special Education, 27*, 453–465.

Gresham, F. M., MacMillan, D. L., & Bocian, K. M. (1996). Learning disabilities, low achievement, and mild mental retardation: More alike than different? *Journal of Learning Disabilities, 29*, 570–581.

Grossman, H. J. (Ed.). (1973). *Manual on terminology and classification in mental retardation.* Washington, D.C.: American Association on Mental Deficiency.

Grossman, H. J. (Ed.). (1983). *Classification in mental retardation* (3rd rev.). Washington, D.C.: American Association on Mental Deficiency.

Gullicksen, H. (1950). *Theory of mental tests.* New York: Wiley.

Hanna, G. S., Dyck, N. A., & Holen, M. C. (1979). Objective analysis of achievement-aptitude discrepancies in LD classification. *Learning Disability Quarterly, 2*, 32–38.

Harris, A. J. (1961). *How to increase reading ability* (4th ed.). New York: McKay.

Harris, A. J. (1971). *How to increase reading ability* (5th ed.). New York: McKay.

Harris, A. J. (1975). *How to increase reading ability* (6th ed.). New York: McKay.

Heber, R. (1961). Modifications in the manual on terminology and classification in mental retardation. *American Journal of Mental Deficiency, 65*, 499–500.

Heber, R. F. (1959). A manual on terminology and classification in mental retardation. *American Journal of Mental Deficiency Monograph* (Supp. 64).

Hempel, C. (1952). Fundamentals of concept formation in empirical science. *International Encyclopedia of Unified Science* (Vol. II, No. 7). Chicago: University of Chicago Press.

Hessler, G. L. (1987). Educational issues surrounding severe discrepancy. *Learning Disabilities Research, 3*, 43–49.

Hinshaw, S. (1992). Externalizing behavior problems and academic underachievement in childhood and adolescence: Causal relationships and underlying mechanisms. *Psychological Bulletin, 111*, 127–155.

Hoffman, J. V. (1980). The disabled reader: Forgive us our regressions and lead us not into expectations. *Journal of Learning Disabilities, 13*, 7–11.

Hollingworth, L. S. (1926). *The psychology of subnormal children.* New York: Macmillan.

Individuals with Disabilities Education Act Amendments of 1997. Public Law 105-17. 11stat. 37 (20 U.S.C. 1401 [26]).

Ingram, C. P. (1935). *The education of the slow-learning child.* New York: Ronald Press.

Jastak, J. F. (1967). Mental Retardation. *Science, 155*, 577–578.

Johnson, D. J., & Myklebust, H. R. (1967). *Learning disabilities: Educational principles and pratices.* New York: Grune & Stratton.

Juel, C. (1988). Learning to read and write: A longitudinal study of 54 children from first through fourth grades. *Journal of Educational Psychology, 78*, 243–245.

Kavale, K. A. (1987). Theoretical issues surrounding severe discrepancy. *Learning Disabilities Research, 3*, 12–20.

Kavale, K. A. (1993). A science and theory of learning disabilities. In: G. R. Lyon, D. B. Gray, J. F. Kavanagh & N. A. Krasnegor (Eds), *Better Understanding Learning Disabilities: New*

Views from Research and Their Implications for Education and Public Policies (pp. 171–195). Baltimore: Brookes.

Kavale, K. A. (1995). Setting the record straight on learning disability and low achievement: The tortuous path of ideology. *Learning Disabilities Research and Practice, 10*, 145–152.

Kavale, K. A., & Forness, S. R. (1984). A meta-analysis of the validity of Wechsler Scale profiles and recategorizations: Patterns or parodies? *Learning Disability Quarterly, 7*, 136–156.

Kavale, K. A., & Forness, S. R. (1985). *The science of learning disabilities*. San Diego, CA: College-Hill Press.

Kavale, K. A., & Forness, S. R. (1994). Learning disabilities and intelligence: An uneasy alliance. In: T. E. Scruggs & M. A. Mastropieri (Eds), *Advances in Learning and Behavioral Disabilities* (Vol. 8, pp. 1–63). Greenwich, CT: JAI Press.

Kavale, K. A., & Forness, S. R. (1995). *The nature of learning disabilities: Critical elements of diagnosis and classification*. Mahwah, NJ: Lawrence Erlbaum.

Kavale, K. A., & Forness, S. R. (2000). What definitions of learning disability say and don't say: A critical analysis. *Journal of Learning Disabilities, 33*, 239–256.

Kavale, K. A., Forness, S. R., & Lorsbach, T. C. (1991). Definition for definitions of learning disabilities. *Learning Disability Quarterly, 14*, 257–266.

Kavale, K. A, Fuchs, D., & Scruggs, T. E. (1994). Setting the record straight on learning disability and low achievement: Implications for policymaking. *Learning Disabilities Research and Practice, 9*, 70–77.

Kavale, K. A, & Reese, J. H. (1992). The character of learning disabilities: An Iowa profile. *Learning Disability Quarterly, 15*, 74–94.

Kaye, D. B., & Baron, M. B. (1987). Long-term stability of intelligence and achievement scores in specific-learning-disabilities samples. *Journal of Psychoeducational Assessment, 3*, 257–266.

Kirk, S. A. (1962). *Educating exceptional children*. Boston: Houghton Mifflin.

Kirk, S. A. (1975). Behavioral diagnosis and remediation of learning disabilities. In: S. A. Kirk & J. M. McCarthy (Eds), *Learning Disabilities: Selected ACLD Papers*. Boston: Houghton Mifflin.

Kirk, S. A., & Elkins, J. (1975). Characteristics of children enrolled in the child service demonstration centers. *Journal of Learning Disabilities, 8*, 630–637.

Klatt, H. J. (1991). Learning disabilities: A questionable construct. *Educational Theory, 41*, 47–60.

Koppitz, E. M. (1971). *Children with learning disabilities: A five year follow-up study*. New York: Grune & Stratton.

Lally, M. J., Lloyd, R. D., & Kulberg, J. M. (1987). Is intelligence stable in learning-disabled children? *Journal of Psychoeducational Assessment, 3*, 411–416.

Lester, G., & Kelman, M. (1997). State disparities in the diagnosis and placement of pupils with learning disabilities. *Journal of Learning Disabilities, 30*, 599–607.

Lloyd, J., Sabatino, D., Miller, T., & Miller, S. (1977). Proposed federal guidelines: Some open questions. *Journal of Learning Disabilities, 10*, 69–71.

Lyon, G. R. (1989). I.Q. is irrelevant to the definition of learning disabilities: A position in search of logic and data. *Journal of Learning Disabilities, 22*, 504–506, 512.

Macmann, G. M., Barnett, D. W., Lombard, T. J., Belton-Kocher, E., & Sharpe, M. N. (1989). On the actuarial classification of children: Fundamental studies of classification agreement. *Journal of Special Education, 23*, 127–143.

MacMillan, D., Siperstein, G., & Gresham, F. (1996). A challenge to the viability of mild mental retardation as a diagnostic category. *Exceptional Children, 62*, 356–371.

MacMillan, D. L., Gresham, F. M., & Bocian, K. M. (1998). Discrepancy between definitions of learning disabilities and school practices: An empirical investigation. *Journal of Learning Disabilities, 31*, 314–326.

MacMillan, D. L., Gresham, F. M., Siperstein, G. N., & Bocian, K. M. (1996). The labyrinth of IDEA: School decisions on referred students with subaverage general intelligence. *American Journal on Mental Retardation, 101*, 161–174.

MacMillan, D. L., & Jones, R. L. (1972). Lions in search of more Christians. *Journal of Special Education, 6*, 81–91.

Macy, M., Baker, J., & Kosinski, S. (1979). An empirical study of the Myklebust learning quotient. *Journal of Learning Disabilities, 12*, 93–96.

Mather, N., & Healey, W. C. (1990). Deposing aptitude-achievement discrepancy as the imperial criterion for learning disabilities. *Learning Disabilities: A Multidisciplinary Journal, 1*, 40–48.

McGill-Franzen, S., & Allington, R. L. (1993). Flunk 'em or get them classified: The contamination of primary grade accountability data. *Educational Researcher, 22*, 19–22.

McKnight, P. T. (1982). The learning disability myth in American education. *Journal of Education, 164*, 351–359.

McLeod, J. (1979). Educational underachievement: Toward a defensible psychometric definition. *Journal of Learning Disabilities, 12*, 322–330.

McLesky, J. (1989). The influence of level of discrepancy on the identification of students with learning disabilities. *Journal of Learning Disabilities, 22*, 435–438, 443.

McLeskey, J. (1992). Students with learning disabilities at primary, intermediate, and secondary grade levels: Identification and characteristics. *Learning Disability Quarterly, 15*, 13–19.

McLeskey, J., & Waldron, N. L. (1990). The identification and characteristics of students with learning disabilities in Indiana. *Learning Disabilities Research, 5*, 72–78.

McLesky, J., & Waldron, N. L. (1991). Identifying students with learning disabilities: The effect of implementing statewide guidelines. *Journal of Learning Disabilities, 24*, 501–506.

McPhail, A. H. (1922). The correlation between the I. Q. and the A. Q. *School and Society, 16*, 586–588.

Meehl, P. E. (1954). *Clinical versus statistical prediction.* Minneapolis: University of Minnesota Press.

Mercer, C. D., Jordan, L., Allsopp, D. H., & Mercer, A. R. (1996). Learning disabilities definitions and criteria used by state education departments. *Journal of Learning Disabilities, 19*, 217–232.

Messick, S. (1981). Constructs and their vicissitudes in educational and psychological measurement. *Psychological Bulletin, 89*, 575–588.

Meyen, E. (1989). Let's not confuse test scores with the substance of the discrepancy model. *Journal of Learning Disabilities, 22*, 482–483.

Monroe, M. (1932). *Children who cannot read.* Chicago: University of Chicago Press.

Morris, R. D., Stuebing, K. K., Fletcher, J. M., Shaywitz, S. E., Lyon, G. R., Shankweiler, D. P., Katz, L., Francis, D. J., & Shaywitz, B. A. (1998). Subtypes of reading disability: Variability around a phonological core. *Journal of Educational Psychology, 90*, 347–373.

Morrison, G. M., MacMillan, D. L., & Kavale, K. A. (1985). System identification of learning disabled children: Implications for research sampling. *Learning Disability Quarterly, 8*, 2–10.

Myklebust, H. R., Bannochie, M. N., & Killen, J. R. (1971). Learning disabilities and cognitive processes. In: H. R. Myklebust (Ed.), *Progress in Learning Disabilities* (Vol. II, pp. 213–251). New York: Grune & Stratton.

Naglieri, J. A. (1981). Factor structure of the WISC-R for children identified as learning disabled. *Psychological Reports, 49*, 891–895.

Noel, M. M., & Fuller, B. C. (1985). The social policy construction of special education: The impact of state characteristics on identification and integration of handicapped children. *Remedial and Special Education, 6*, 27–35.

Norman, C. A., & Zigmond, N. (1980). Characteristics of children labeled and served as learning disabled in school systems affiliated with Child Service Demonstration Centers. *Journal of Learning Disabilities, 13*, 542–547.

O'Donnell, L. E. (1980). Intra-individual discrepancy in diagnosing specific learning disabilities. *Learning Disability Quarterly, 3*, 10–18.

O'Shea, L. J., & Velcaute, G. (1986). A comparison over time of relative discrepancy scores of low achievers. *Exceptional Children, 53*, 253–259.

Payne, R. W., & Jones, H. G. (1957). Statistics for the investigation of individual cases. *Journal of Clinical Psychology, 13*, 115–121.

Pennington, B. F., Gilger, J. W., Olson, R. K., & DeFries, J. C. (1992). The external validity of age-versus I.Q.-discrepancy definitions of reading disability: Lessons from a twin study. *Journal of Learning Disabilities, 25*, 562–573.

Perlmutter, B., & Perus, M. (1983). Identifying children with learning disabilities: A comparison of diagnostic procedures across school districts. *Learning Disability Quarterly, 6*, 321–328.

Pintner, R., & Marshall, H. (1921). A combined mental-educational survey. *Journal of Educational Psychology, 12*, 32–43.

Piotrowski, R. J., & Siegel, D. J. (1986). The I.Q. of learning disability samples: A re-examination. *Journal of Learning Disabilities, 19*, 492–493.

Popenoe, H. (1927). A report of certain significant deficiencies of the accomplishment quotient. *Journal of Educational Research, 16*, 40–47.

Public Law 94–142. (1975, November 29). *Education for All Handicapped Children Act*. Washington, D.C.: U.S. Congress.

Reger, R. (1979). Learning disabilities: Futile attempts at a simplistic definition. *Journal of Learning Disabilities, 12*, 31–34.

Reschly, D. J. (1992). Mental retardation: Conceptual foundation, definitional criteria, and diagnostic operations. In: S. R. Hooper, G. W. Hynd & R. E. Mattison (Eds), *Developmental Disorders: Diagnostic Criteria and Clinical Assessment* (pp. 23–67). Hillsdale, NJ: Lawrence Erlbaum.

Reynolds, C. R. (1981). The fallacy of "two years below grade level for age" as a diagnostic criterion for reading disorders. *Journal of School Psychology, 19*, 350–358.

Reynolds, C. R. (1984–1985). Critical measurement issues in learning disabilities. *Journal of Special Education, 18*, 451–476.

Reynolds, C. R. (1985). Measuring the aptitude-achievement discrepancy in learning disability diagnosis. *Remedial and Special Education, 6*, 37–55.

Reynolds, C. R. (1986). Toward objective diagnosis of learning disabilities. *Special Services in the Schools, 2*, 161–176.

Rispens, J., van Yperen, T. A., & van Duijn, G. A. (1991). The irrelevance of I.Q. to the definition of learning disabilities: Some empirical evidence. *Journal of Learning Disabilities, 24*, 434–438.

Rodenborn, L. V. (1974). Determining and using expectancy formulas. *The Reading Teacher, 28*, 286–291.

Rodgers, B. (1983). The identification and prevalence of specific reading retardation. *British Journal of Educational Psychology, 53*, 369–373.

Ruch, G. M. (1923). The achievement quotient technique. *Journal of Educational Psychology, 14*, 334–343.

Rutter, M. (1978). Prevalence and types of dyslexia. In: A. L. Benton & D. Pearl (Eds), *Dyslexia: An Appraisal of Current Knowledge* (pp. 5–28). New York: Oxford University Press.

Rutter, M. (1990). Isle of Wight revisited: Twenty-five years of child psychiatric epidemiology. *Journal of American Academy of Child and Adolescent Psychiatry, 29*, 633–653.

Rutter, M., & Yule, W. (1973). Specific reading retardation. In: L. Mann & D. Sabatino (Eds), *The First Review of Special Education* (pp. 49–62). Philadelphia: JSE Press.

Rutter, M., & Yule, W. (1975). The concept of specific reading retardation. *Journal of Child Psychology and Psychiatry, 16*, 181–197.

Sabatino, D. A., & Miller, T. L. (1980). The dilemma of diagnosis in learning disabilities: Problems and potential directions. *Psychology in the Schools, 17*, 76–86.

Salvia, J. & Clark, J. (1973). Use of deficits to identify the learning disabled. *Exceptional Children, 39*, 305–308.

Schulte, A. & Borich, G. D. (1984). Considerations in the use of difference scores to identify learning-disabled children. *Journal of School Psychology, 22*, 381–390.

Scruggs, T. E. (1987). Theoretical issues surrounding severe discrepancy: A discussion. *Learning Disabilities Research, 3*, 21–23.

Senf, G. M. (1977). A perspective on the definition of LD. *Journal of Learning Disabilities, 10*, 537–539.

Share, D. L., McGee, R., McKenzie, D., Williams, S., & Silva, P. A. (1987). Further evidence relating to the distinction between specific reading retardation and general reading backwardness. *British Journal of Developmental Psychology, 5*, 34–44.

Share, D. L., McGee, R., & Silva, P. D. (1989). I.Q. and reading progress: A test of the capacity notion of I.Q. *Journal of American Academy of Child and Adolescent Psychiatry, 28*, 97–100.

Share, D. L., & Silva, P. A. (1986). The stability and classification of specific reading retardation: A longitudinal study from age 7 to 11. *British Journal of Educational Psychology, 56*, 32–39.

Share, D. L., & Silva, P. A. (1987). Language deficits and specific reading retardation: Cause or effect? *British Journal of Disorders of Communication, 22*, 219–226.

Shaw, S. F., Cullen, J. P., McGuire, J. M., & Brinckerhoff, L. C. (1995). Operationalizing a definition of learning disabilities. *Journal of Learning Disabilities, 28*, 586–597.

Shaywitz, B. A., Fletcher, J. M., Holahan, J. M., & Shaywitz, S. A. (1992). Discrepancy compared to low achievement definitions of reading disability: Results from the Connecticut Longitudinal Study. *Journal of Learning Disabilities, 25*, 639–648.

Shaywitz, B. A., Holford, T. R., Holohan, J. M., Fletcher, J. M., Stuebing, K. K., Francis, D. J., & Shaywitz, S. E. (1995). A Matthew effect for I.Q. but not for reading: Results from a longitudinal study. *Reading Research Quarterly, 30*, 894–906.

Shaywitz, S. E., Escobar, M. D., Shaywitz, B. A., Fletcher, J. M., & Makuch, R. (1992). Evidence that dyslexia may represent the lower tail of a normal distribution of reading ability. *New England Journal of Medicine, 326*, 145–150.

Shaywitz, S. E., Shaywitz, B. A., Fletcher, J. M., & Escobar, M. D. (1990). Prevalence of reading disability in boys and girls: Results of the Connecticut Longitudinal Study. *Journal of the American Medical Association, 264*, 998–1002.

Shepard, L. (1980). An evaluation of the regression discrepancy method for identifying children with learning disabilities. *Journal of Special Education, 14*, 79–91.

Shepard, L. A. (1983). The role of measurement in educational policy: Lessons from the identification of learning disabilities. *Educational Measurement: Issues and Practices, 2*, 4–8.

Shepard, L. A. (1989). Identification of mild handicaps. In. R. Linn (Ed.), *Educational measurement* (3rd ed., pp. 545–572). New York: Macmillan.

Shepard, L. A., & Smith, M. L. (1983). An evaluation of the identification of learning disabled students in Colorado. *Learning Disability Quarterly, 6*, 115–127.

Shepard, L. A., Smith, M. L., & Vojir, C. P. (1983). Characteristics of pupils identified as learning disabled. *American Educational Research Journal, 20*, 309–331.

Short, E. J., Feagans, L., McKinney, J. D., & Appelbaum, M. I. (1986). Longitudinal stability of LD subtypes based on age- and I.Q.-achievement discrepancies. *Learning Disability Quarterly, 9*, 214–225.

Siegel, E. (1968). Learning disabilities: Substance or shadow. *Exceptional Children, 35*, 433–437.

Siegel, L. S. (1988). Evidence that I.Q. scores are irrelevant to the definition and analysis of reading disability. *Canadian Journal of Psychology, 42*, 201–215.

Siegel, L. S. (1989). I.Q. is irrelevant to the definition of learning disabilities. *Journal of Learning Disabilities, 22*, 469–478, 486.

Siegel, L. S. (1992). An evaluation of the discrepancy definition of dyslexia. *Journal of Learning Disabilities, 25*, 618–629.

Simmons, G. A., & Shapiro, B. J. (1968). Reading expectancy formulas: A warning note. *Journal of Reading, 2*, 625–629.

Sinclair, E., & Alexson, J. (1986). Learning disability discrepancy formulas: Similarities and differences among them. *Learning Disabilities Research, 1*, 112–118.

Singer, J., Palfrey, J., Butter, J., & Walker, D. (1989). Variation in special education classification across school districts: How does where you live affect what you are labeled? *American Educational Research Journal, 26*, 261–281.

Smith, M. D., Coleman, J. M., Dokecki, P. R., & Davis, E. E. (1977). Intellectual characteristics of school labeled learning disabled children. *Exceptional Children, 43*, 352–357.

Sobel, M., & Kelemen, E. (1984). On an application of statistical ranking and selection tables to the identification of the learning disabled. *Journal of Educational Statistics, 9*, 215–226.

Stanovich, K. E. (1986). Cognitive processes and the reading problems of learning disabled children: Evaluating the assumption of specificity. In: J. K. Torgesen & B. Y. L. Wong (Eds), *Psychological and Educational Perspectives on Learning Disabilities* (pp. 87–131). Orlando, FL: Academic Press.

Stanovich, K. E. (1986). Matthew effects in reading: Some consequences of individual differences in the acquisition of literacy. *Reading Research Quarterly, 21*, 360–407.

Stanovich, K. E. (1989). Has the learning disabilities field lost its intelligence? *Journal of Learning Disabilities, 22*, 487–492.

Stanovich, K. E. (1991a). Conceptual and empirical problems with discrepancy definitions of reading disability. *Learning Disability Quarterly, 14*, 269–280.

Stanovich, K. E. (1991b). Discrepancy definitions of reading disability: Has intelligence led us astray? *Reading Research Quarterly, 26*, 7–29.

Stanovich, K. E., & Siegel, L. S. (1994). Phenotypic performance profile of children with reading disabilities: A regression-based test of the phonological-core variable-difference model. *Journal of Educational Psychology, 86*, 24–53.

Stevenson, J. (1988). Which aspects of reading disability show a "hump" in their distribution? *Applied Cognitive Psychology, 2*, 77–85.

Sulzbacher, S., & Kenowitz, L. A. (1977). At last, a definition of learning disabilities we can live with? *Journal of Learning Disabilities, 10*, 67–69.

Swanson, H. L. (1991). Operational definitions and learning disabilities: An overview. *Learning Disability Quarterly, 14*, 242–254.

Swerling, S. L., & Sternberg, R. J. (1996). *Off track: When poor readers become learning disabled*. Boulder, CO: Westview Press.

Tarjan, G., Wright, S. W., Eyman, R. K., & Keeran, C. V. (1975). Natural history of mental retardation: Some aspects of epidemiology. *American Journal of Mental Deficiency, 77*, 369–379.

Thorndike, R. L. (1963). *The concepts of over- and underachievement*. New York: Teachers College, Columbia University Press.

Thurlow, M. L., Ysseldyke, J. E., & Casey, A. (1984). Teachers' perceptions of criteria for identifying learning disabled students. *Psychology in the Schools, 21*, 349–355.

Thomson, M. (1982). Assessing the intelligence of dyslexic children. *Bulletin of the British Psychological Society, 35*, 94–96.

Toops, H. A., & Symonds, P. M. (1922). What shall be expected of the A.Q? *Journal of Educational Psychology, 13*, 513–528.

U.S. Department of Education. (1999). *To assure the free appropriate public education of all children with disabilities: Twenty-first annual report to the Congress on the implementation of The Individuals with Disabilities Education Act*. Washington, D.C.: Author.

U.S. Office of Education. (1968). *First annual report of the National Advisory Committee on Handicapped Children*. Washington, D.C.: U.S. Department of Health, Education, and Welfare.

U.S. Office of Education (1976, December 29). Proposed rule-making. *Federal Register, 41*, (230), 52404–52407. Washington, D.C.: U.S. Government Printing Office.

U.S. Office of Education (1977, December 29). Assistance to states for education of handicapped children: Procedures for evaluating specific learning disabilities. *Federal Register, 42*, (250), 65082–65085. Washington, D.C.: U.S. Government Printing Office.

Valus, A. (1986). Achievement-potential discrepancy status of students in LD programs. *Learning Disability Quarterly, 9*, 200–205.

Valus, A. (1986). Teacher perceptions of identification criteria emphasized in initial learning disabilities placement. *Learning Disabilities Research, 2*, 21–25.

Van den Bos, K. P. (1989). Relationship between cognitive development, decoding skill, and reading comprehension in learning disabled Dutch children. In: P. Aaron & M. Josh (Eds), *Reading and Writing Disorders in Different Orthographic Systems* (pp. 75–86). Dordrecht, The Netherlands: Kluwer.

Van der Wissel, A., & Zegers, F. E. (1985). Reading retardation revisited. *British Journal of Developmental Psychology, 3*, 3–9.

Vellutino, F. R., Scanlon, D. M., Sipay, E. R., Small, S. G., Pratt, A., Chen, R., & Denckla, M. B. (1996). Cognitive profiles of difficult-to-remediate and readily remediated poor readers: Early intervention as a vehicle for distinguishing between cognitive and experiential deficits as basic causes of specific reading disability. *Journal of Educational Psychology, 88*, 601–638.

Walberg, H. J., & Tsai, S. (1983). Matthew effects in education. *American Educational Research Journal, 20*, 359–373.

Weir, M. (1967). Mental Retardation. *Science, 155*, 576–577.

Whipple, G. M. (1922). Educational determinism: A discussion of Professor Bagley's address at Chicago. *School and Society, 15*, 599–602.

White, W. J., & Wigle, S. E. (1986). Patterns of discrepancy over time as revealed by a standard-score comparison formula. *Learning Disabilities Research, 2*, 14–20.

Wilson, J. D., & Spangler, P. F. (1974). The Peabody Individual Achievement Test as a clinical tool. *Journal of Learning Disabilities, 7*, 384–387.

Wilson, L., Cone, T., Bradley, C., & Reese, J. (1986). The characteristics of learning disabled and other handicapped students referred for evaluation in the state of Iowa. *Journal of Learning Disabilities, 19*, 553–557.

Wilson, L. R. (1985). Large-scale learning disability identification: The reprieve of a concept. *Exceptional Children, 52*, 44–51.

Wilson, L. R., & Cone, T. (1984). The regression equation method of determining academic discrepancy. *Journal of School Psychology, 22*, 95–110.

Wilson, L. R., Cone, T. E., Busch, R., & Allee, T. (1983). A critique of the expectancy formula approach: Beating a dead horse? *Psychology in the Schools, 20*, 241–249.

Wilson, W. R. (1928). The misleading accomplishment quotient. *Journal of Educational Research, 17*, 1–10.

Willson, V. L. (1987). Statistical and psychometric issues surrounding severe discrepancy. *Learning Disabilities Research, 3*, 24–28.

Ysseldyke, J., Algozzine, B., & Epps, S. (1983). A logical and empirical analysis of current practice in classifying students as handicapped. *Exceptional Children, 50*, 160–166.

Ysseldyke, J. E., Algozzine, B., Richey, L., & Graden, J. L. (1982). Declaring students eligible for learning disability service: Why bother with the data? *Learning Disability Quarterly, 5*, 37–44.

Ysseldyke, J. E., Algozzine, B., Shinn, M. R., & McGue, M. (1982). Similarities and differences between low achievers and students classified learning disabled. *Journal of Special Education, 16*, 73–85.

Ysseldyke, J. E., Christenson, S., Pianta, B., & Algozzine, B. (1983). An analysis of teachers' reasons and desired outcomes for students referred for psychoeducational assessment. *Journal of Psychoeducational Assessment, 1*, 73–83.

Yule, W. (1973). Differential prognosis of reading backwardness and specific retardation. *British Journal of Educational Psychology, 43*, 244–248.

Yule, W., & Rutter, M. (1976). Epidemiology and social implications of specific reading retardation. In: R. M. Knights & D. J. Bakker (Eds), *The Neuropsychology of Learning Disorders: Theoretical Approaches* (pp. 25–39). Baltimore: University Park Press.

Zigler, E. (1967). Familial mental retardation: A continuing dilemma. *Science, 155*, 292–298.

Zigler, E., & Balla, D. (Eds). (1982). *Mental retardation: The developmental-difference controversy.* Hillsdale, NJ: Lawrence Erlbaum.

Zigler, E., & Hodapp, R. (1986). *Understanding mental retardation.* New York: Cambridge University Press.

STRATEGIC COHERENCE AND ACADEMIC ACHIEVEMENT

Angelica Moè, Cesare Cornoldi and Rossana De Beni

ABSTRACT

Unsuccessful students stand out in many ways, one of which is their lower efficacy in study methods. In the present research we hypothesized that low academic achievement is accompanied by a low 'strategic coherence', i.e. by a poor capacity to use the strategies which are considered the most effective to study a text. Study 1 found that low achievement undergraduates have lower strategic coherence than high achievers. Study 2 demonstrated that the less coherent students have a poor study method and low scholastic performance. The paper concludes with a discussion on the importance of strategic coherence for an effective studying.

Academic achievement depends on numerous factors related to study methods and learning strategies as well as different levels of intellectual efficacy, or the absence of specific learning problems. Among these, recent research has demonstrated the importance of good study habits and use of strategies (Hettich, 1998; McCutchen, Francis & Kerr, 1997; Pressley, Van Etten, Yokoi, Freebern & Van Meter, 1998; Pressley, Yokoi, Van Etten & Freebern, 1997; Van Etten, Freebern & Pressley, 1997; Weinstein & Hume, 1998), previous knowledge of the material (McCutchen et al., 1997; Pressley et al., 1998), the type of text and quality of teaching (Pressley et al., 1997), motivation to learn and success (Van Etten et al., 1997; Van Etten, Pressley, Freebern & Eschevarria, 1998) and cognitive style (Beishiuzen & Stoutjesdijk, 1999).

Technological Applications, Volume 15, pages 237–258.

ISBN: 0-7623-0815-x

Research contrasting the study method and study strategies of high and low achievers offers important information concerning both the differences between the two groups and the different effectiveness of various study strategies. In fact, high achievers are distinguished from low achievers not only for their greater use of strategies in general (e.g. Stoynoff, 1996) but also for the type of strategy prevalently used. High achievers prefer schema-driven study methods to text-driven, which on the one hand may require greater cognitive resources but on the other permit them to adapt the contents, elaborating them according to their own cognitive schemes and effective strategies such as self-testing (Wood, Motz & Willoughby, 1998) and note-taking (Loranger, 1994). They make little use of simple underlining or read-and-repeat techniques (Wood et al., 1998) which are widely used by low achievers and are characterized by strict adherence to the text and reduced personal elaboration of the contents (Gadzella, 1995; Turner, 1992). High achievers have a greater organizational ability (Kleijn, van der Ploeg & Topman, 1994), are able to accurately distinguish between main ideas and details (Stoynoff, 1997), have a study approach which is distributed in time, so that particular attention can be dedicated to revision and means of concentrating better (Wilding & Valentine, 1992), they are more flexible in choice of strategy according to the subject being studied (Wood et al., 1998) and are more motivated to learn (Albaili, 1997; Loranger, 1994). On the other hand low achievers have greater difficulty managing independently and organizing their study. They are unable to balance the different subjects to be prepared or the different phases of content acquisition, so that certain aspects, such as revision, are often ignored (Wilding & Valentine, 1992). This leads them to face learning tasks rather passively, without motivation and with little reflection and planning (Herrmann, Raybeck & Gutman, 1993).

Despite the fact that research has analyzed such a large variety of strategic differences between high and low achievers, only a few studies have examined the students metacognitive theories associated to low achievers' study activity. Furthermore, results of these studies appear ambiguous. For example, it has been suggested that low achievers are unaware of the inefficiency of their study method (Loranger, 1994). However, this low self-knowledge could reflect an absence of self-evaluation and self-control abilities, rather than a lack of the capacity to recognize the importance of good strategies. In fact, it has been observed (see for example Wolters, 1998) that low achievers have a typically low perception of self-efficacy and therefore tend to consider themselves incapable of setting up the necessary processes which lead to good results in learning. This low perception of self-efficacy does not seem to correspond to a state of "blissful ignorance" by which an individual would use bad strategies but at the same time consider them appropriate and efficient. On the contrary, a low perception of self-efficacy could be related to an intuition that good strategies do exist, but are not those used.

To examine whether low achievers are able to recognize the importance of good study strategies, it is necessary to know and measure not only the low achievers' use of these strategies, but also how valid the strategies are considered by the same students independently of their effective use. Student use ratings should refer to the perception of the self as a student (actual studying) and therefore to a real self. The student efficacy ratings, on the other hand, should describe metacognitive beliefs about studying and refer to the ideal perception of the self as a student. Therefore, the efficacy ratings could be considered as an index of a particular subsystem of the students' ideal self (describing how the student would like to study), or rather of the imperative self (describing how the student believes he/she should study) (Higgins, 1987; Markus & Nurius, 1986).

Opposing, actual and ideal, representations of the self, and the possible discrepancy between the two, originally examined in social and personality clinical contexts, have been considered more recently in learning situations in reference to the levels that a student retains she/he is able to reach through study, assuming important motivational sources (e.g. Leondari, Syngollitou & Kiosseoglou, 1998; Oyserman, Gant & Ager, 1995). The same perspective, applied to study methods, suggests that low achievers, when faced with a series of different study strategies, should consider strategies they do not use in their actual study as most effective and representative of an ideal student and, on the contrary, that they should report that they use strategies they consider less effective. The present research focused on this aspect.

In the present work, the efficacy and use ratings have been compared with the aim of obtaining an index of the extent that the student thinks she/he uses the strategies which are considered most effective and representative of the ideal way to study. We have defined this relation between strategies appreciated and those used with the term 'strategic coherence'. Therefore, strategic coherence is produced by a correspondence between the effectiveness ratings assigned to study strategies and use ratings referred to the degree to which the student thinks she/he uses these strategies. A smaller distance reflects greater strategic coherence while greater distances indicate little strategic coherence.

Generally, self-evaluation questionnaires regarding method and/or study strategies require only a use rating and not an efficacy rating. We know of only three studies which have required an efficacy rating of the strategies. Nolen (1988) has considered two efficacy ratings, one general and the other specific to the subject, and a use rating, finding important relations between motivational aspects and use and efficacy ratings. Huet and Mariné (1996) asked subjects (mean age 18.2 years) to evaluate the effectiveness of three strategies and then measured their actual use in a memory task. In both cases the efficacy and use ratings were not compared directly, nor were the effects on performance due to the coherence between efficacy and use ratings considered.

The importance of evaluating the implications of strategic coherence has been suggested for the first time by De Beni and Moè (1997) who examined low achieving university students who came to the University Psychology Service because of their study difficulties. It turned out that they knew extremely valid study strategies, but did not apply them. Instead they preferred to study using means that they too considered less adequate, but had become habit over time and difficult to modify. In particular they did not use any demanding strategies, such as revising some time later, the use of mnemonics, drawing up tables and diagrams, simulating exam questions or developing personal ideas. Training was proposed to students to make them more aware of the fact that the study strategies they adopted were not those generally considered effective, and therefore of little strategic coherence. It was associated with a reflection about the reasons why this occurs and a teaching of more effective study methods allowing the students to gain better results in exams and have more confidence in themselves and a greater motivation to study.

In this work, we wanted to systematically test the hypothesis that an important characteristic differentiating low and high achievers could be strategic coherence. Lower academic achievement should correspond to lower strategic coherence. Students who are less coherent shall be aware of studying (actual studying) not coherently with their ideal way (belief about studying). We think this could reflect not only the selection of appropriate strategies, but also lower levels of metacognitive control, and probably also of self-efficacy, which are both critical to good methodology and motivation to learn. This hypothesis has been opposed by the hypothesis of "blissful ignorance" by which bad students are satisfied with their study methods and use inappropriate strategies, thinking them adequate.

This research was therefore aimed at examining the role of strategic coherence on study methods and scholastic performance. The first study was carried out with the aim of verifying the hypothesis that low achievers differ from high achievers in their lower strategic coherence. The second study was aimed at studying the characteristics of more or less coherent students by evaluating their study methods and scholastic performance with the hypothesis that greater strategic coherence corresponds to both better study methods and better scholastic performance.

STUDY 1

Study 1 was aimed at studying a group of low achievers for strategic coherence in declared strategy effectiveness and use. The hypothesis was tested on the basis of the Strategy Questionnaire proposed by Cornoldi (1995). The Questionnaire presents 39 cognitive and behavioral study strategies which are very often reported by students of different ages. On the basis of the available literature, some of these strategies appear particularly effective, whereas for others the effectiveness is not evident. If low achievers are aware of the different effectiveness of different study strategies their efficacy ratings would not be significantly different from those given by a control group of high achievers. Low achievers' study method could be inadequate because of lower use of the strategies they themselves consider as good. If, on the contrary, low achievers are unaware of good study strategies, clear differences between them and high achievers should also be evident at the metacognitive level of efficacy ratings.

In the present study, the measure of students' success was considered as the number of examinations passed in the first year at an Italian University (corresponding approximately to the first year of undergraduate studies). The number of examinations passed is important in the Italian context as students are free to define examination times and, furthermore, success in first year is a powerful predictor of ensuing academic career. More than 20% of first year Italian students drop out of university mainly because they are unable to pass a sufficient number of examinations (Bernardi & Valente, 1998).

Method

Participants

19 low achievers (4 males and 15 females) (mean age 19 years), participated in the study. Subjects were selected on the basis of the low number – either one or two – of exams passed. Low achievers were compared with a control group of 14 high achievers of the same age who had satisfactorily met University requirements, having passed six or all seven first year examinations.

Materials and Procedure

We used the Strategy Questionnaire proposed by Cornoldi (1995), with slight modifications (see Appendix 1). The Strategy Questionnaire presents a list of

39 study strategies concerning different aspects of study for an examination such as organization, comprehension, underlining, note-taking, memorizing, reviewing and anticipating the exam context. Examples of items presumably associated with effective strategies are as follows: "write guiding concepts next to the text", "review material after an extended period of time." Examples of items associated with strategies whose effects on achievement are not always relevant and could sometimes be even negative are as follows: "read the text aloud", "interpolate the study of other issues, in the study of a chapter". The Strategy Questionnaire was administered in a pilot study to a sample of 113 students attending the second year of University, obtaining Cronbach's alphas for efficacy ratings, use ratings and strategic coherence of 0.82, 0.82, 0.72 respectively.

The Strategy Questionnaire was administered in two phases. In the first phase students were invited to rate the efficacy of each strategy, irrespective of their actual use. They were required to use a 7-point Likert-type scale, where 1 corresponded to the lowest efficacy and 7 to the highest. In particular, the instructions asked students to "consider the ideal case of an individual who, within reasonable time limits, calmly follows a good study method for an examination." In order to obtain use ratings which reflected the same context as the efficacy ratings, a second phase followed about forty minutes after, without any session interruption. In the second phase, the same strategies were unexpectedly presented again, but in a different order. The students had to use the same Likert-type scale to rate the use of each strategy, with 1 corresponding to 'no use at all' and 7 to 'constant use'.

The Strategy Questionnaire was administered collectively. During the interval between the two administrations students remained seated and filled in other questionnaires, concerning personal data.

Results

We first compared high and low achievers for academic career. The measure represented by the number of first year examinations passed was in correspondence with the other indices. In particular high achievers were significantly better than low achievers regarding their diploma and mean examination marks at first year university (see Table 1).

Data analysis concerning the Strategy Questionnaire was based on participants' overall ratings regarding the coherence between ratings, and on the different ratings given to selected groups of good and poor strategies, or single strategies.

Table 1. Mean achievement scores obtained by low and high achievers at the end of secondary school (diploma mark) and during their first University year (mean examination mark). Mean overall efficacy and use ratings assigned by them to the 39 strategies presented in the Strategy Questionnaire and mean strategic coherence score (*M* = mean score, *SD* = standard deviation). Student's *t* comparisons between the means of the two groups are reported.

	Low achievers		High achievers		
Variable	*M*	*SD*	*M*	*SD*	$t(31)$
Diploma mark	45.16	6.09	51.86	6.35	3.07 $p = 0.004$
Mean examination mark	25.21	1.87	27.06	1.17	3.27 $p = 0.003$
Efficacy score	175.00	25.98	187.29	19.24	n.s.
Use score	163.68	28.08	171.86	22.50	n.s.
Strategic coherence	45.63	13.16	36.00	10.57	2.25 $p = 0.032$

We first computed three scores on the basis of students' responses to the Strategy Questionnaire. The efficacy rating score was based on the sum of efficacy ratings for the 39 strategies and reflected a general value assigned by students to the efficacy of strategies. The use rating score was based on the sum of use ratings for the same strategies and reflected the general tendency students reported for use strategies. The strategic coherence score was determined by the sum of absolute values of the differences between efficacy and use ratings for each strategy. It must be noted that the absolute values were used because a lack of coherence could be due to a use rating lower than efficacy rating, for each strategy, but could also be due to a use rating greater than the efficacy rating. For example, if a student gave an efficacy rating of '6' and then a use rating of '3' to the same strategy the difference would be '3', but the difference would also be '3' if an efficacy rating of '3' and use rating of '6' had been assigned to the strategy. Consequently low scores in the "strategic coherence" measure reflected a smaller difference between the two ratings and then a higher strategic coherence. Strategic coherence could also be computed by calculating the correlation between efficacy and use rating given to the same strategies for each subject. However, we decided to use the difference procedure for two reasons. The first was that correlation would underestimate the lack of coherence found in the common case where students report high, but differentiated, efficacy ratings for many strategies and lower, but again differentiated, ratings for the same strategies. The second reason was associated with our goal of creating a procedure which could easily be repeated by teachers (and understood by students during metacognitive training) who should find more confidence in computing differences rather than a correlation.

High achievers showed better strategic coherence than low achievers as their score was significantly lower, Student's $t\,(31) = 2.25, p < 0.05$. It is important to note that the two groups presented slight but not significant differences in efficacy and use scores (see Table 1). This result may seem paradoxical, as one would expect a lower coherence to reflect lower use ratings. However, it must be considered that the opposite was also possible, i.e. a low efficacy rating and a higher use rating, thus increasing the coherence score, but decreasing the summed use score.

The general pattern of data is confirmed by a more detailed analysis concerning selected groups of strategies. In fact we selected the 8 strategies we considered most effective and the 8 we considered less effective from the Questionnaire (the corresponding items are marked differently in the Questionnaire presented in Appendix 1). Table 2 presents the mean efficacy ratings and mean use ratings assigned by participants to the two groups of strategies.

As can be seen in Table 2, the two groups presented a similar pattern of efficacy ratings and both gave higher ratings to the most effective strategies. A 2×2 ANOVA for a mixed design showed that this difference was significant,

Table 2. Mean use and efficacy ratings given by low and high achievers to a selected group of 8 good and 8 less effective strategies and mean coherence scores of the two groups for the same strategies (*M* = mean score, *SD* = standard deviation).

	Low achievers		High achievers	
	M	*SD*	*M*	*SD*
Efficacy ratings				
Less effective strategies	30.37	5.01	33.36	4.43
Good strategies	44.21	6.32	46.79	5.82
Use ratings				
Less effective strategies	28.26	6.53	29.36	5.50
Good strategies	40.63	8.74	44.71	5.55
Strategic coherence scores				
Less effective strategies	9.16	4.11	7.57	2.95
Good strategies	10.53	4.22	4.93	1.82

$F(1, 31) = 168.03$, $p < 0.001$, whereas neither group effect nor interaction were significant. A similar result was obtained with the ANOVA concerning use ratings, i.e. the effect due to the quality of strategies was highly significant, $F(1, 31) = 113.41$, $p < 0.001$. Although the high achievers reported a slightly higher use of effective strategies, the interaction between groups and strategies did not reach significance. This result is interesting because it suggests that a discrepancy between efficacy and use ratings of good strategies is not be generally valid, but must be referred to the single subjects, as with a coherence analysis. In fact, Table 2 (bottom) shows that differences between groups were particularly evident in the strategic coherence scores. Low achievers were less coherent (had higher difference scores) for both less effective and highly effective strategies, as is confirmed by a significant group effect, $F(1, 31) = 17.02$, $p < 0.001$. Furthermore, the difference was higher in the case of good strategies (coherence of high achievers was more than double) for less effective strategies. This differential effect was confirmed by the significant interaction between groups and the quality of strategies, $F(1, 31) = 4.99$, $p < 0.05$.

Discussion

The results show that school achievement defined by the number of examinations passed corresponds to both other measures of school achievement. Actually the ability to pass university examinations and follow a regular course of studies seems critical for the student's university life. This ability was also associated with a better preceding carrier (diploma mark) and better marks in the examinations passed.

Furthermore, low achievers obtained a significantly higher score in the differential overall coherence measure showing their lower strategic coherence. This difference could be simply due to the fact their efficacy ratings are similar to those given by high achievers, however they give lower use ratings. However, the differences between the two groups were unclear regarding not only the measure of efficacy, but also the measure of use. This suggests that the difference in strategic coherence is not necessarily due to other measures i. e. to either inability to recognize the importance of study strategies or a declared lower use of them. The result is also confirmed by the fact that the strategic coherence score did not significantly correlate with either the efficacy score (Pearson's $r = -0.07$), or the use score (Pearson's $r = 0.02$).

The analysis concerning selected groups of strategies offered important information. In fact the lower strategic coherence of low achievers could be due to the simple fact that they considered effective the strategies which are, by general agreement, most effective, but did not use them. In this case, the difference between high and low achievers should be reflected by lower use ratings assigned by low achievers to these good strategies. This was not the case however, suggesting the importance of considering not only group differences in efficacy and use ratings, but also individual differences concerning idiosyncratic differences for specific strategies. In other words, despite a general common tendency to give higher efficacy and use ratings to better strategies, each low achiever could have assigned, according to his/her own experience and cognitive profile, specific ratings to different strategies.

The effect concerning coherence scores appears to be due to low achievers more often reporting different ratings for efficacy and use of single strategies. This effect does not seem to be related to a greater effort by high achievers to create congruence between the two ratings. Instructions stressed the importance of giving independent ratings and subjects confirmed that they had met the requirements. However it is possible that high achievers based their efficacy ratings on their own experience, convinced that their experience reflected the best method of study. By contrast, due to academic failure, low achievers could rate as less effective the strategies they knew they used. On the basis of data of this study it was possible to know whether lower strategic coherence is a factor producing lower achievement or a consequence of it. At the same time, even if we assume that strategic coherence may affect academic achievement, we do not think that its role is either unique or central, as a large body of literature shows that other variables, like intellectual ability, prior knowledge, motivation, and so on, may also affect achievement, probably to a larger extent.

Furthermore the difference between groups in strategic coherence concerned good strategies to a greater extent. Low achievers were comparatively more coherent in less effective cognitive and behavioral strategies which are more repetitive, less oriented towards deep elaboration of a text and more towards a search for external support. These latter strategies may not lead to success in university examinations.

In conclusion, Study 1 showed that strategic coherence, as we have defined it, is an important aspect of study, discriminating between high and low achievers attending their first year of university. In order to explore the role of strategic coherence better we decided to follow a different perspective in a second study, i.e. to directly analyze characteristics of students with high or low strategic coherence scores. This involved testing groups of students who, after senior high school (scuola media superiore), are about to begin university. Groups of students were selected on the basis of their strategic coherence score which became the criterion variable for group formation.

STUDY 2

Study 2 tested the hypotheses that students with low strategic coherence have a poorer study method and lower school achievement than those with high strategic coherence. For this purpose we selected specific groups of students characterized by either high or low strategic coherence score from a larger sample of students. These students were administered a Questionnaire examining aspects of study method and habits. The Questionnaire was based on the only standardized Study Questionnaire available in Italy and examined five critical aspects: Organization, Elaboration, Prediction, Test Anticipation and Reflection.

We predicted that students who are more coherent are more able to develop an efficient study method, including an ability to organize their personal work (Organization), a deeper personal elaboration of the material (Elaboration), an ability to evaluate task difficulty correctly (Prediction), a capacity to consider the best way to prepare for an examination (Test Anticipation) and the inclination to reflect on their cognitive functioning (Reflection). We also obtained an achievement measure which was represented by a mark in Language, a summed mark indicating achievement in writing and studying Italian literature. In Italian schools all students partake in the Language area to approximately the same degree of difficulty and it represents a large proportion of school activity. We predicted that less coherent students have a poorer study method and, because of their inefficient study method, also lower achievement.

Method

Participants

100 fourth-year senior high school students following different types of curricula were selected from a sample of 234 students, mean age 18.30 years, for having obtained either the highest or lowest strategic coherence scores following the same procedure as in the preceding Study. For this purpose, we administered the same Strategy Questionnaire (Cornoldi, 1995) used in Study 1 to student groups in the classroom, during their normal school activity. In the first phase they were asked to give efficacy ratings and in the second, after a break of forty minutes, to give use ratings. We then computed coherence scores using the same methodology as in Study 1. Forty-nine students (22 males and 27 females) were classified as high-coherent (score below 35), 51 (24 males and 27 females) as low-coherent (score above 60).

Materials and Procedure

We created a Study Method Questionnaire by selecting and adapting five areas of the Study Questionnaire QS proposed by Cornoldi, De Beni and the MT Group (1993), whose psychometric properties are quite good (for a review see Terreni & Campiotti, 1999). The five areas examined the following aspects of study method: Organization, Elaboration, Prediction, Test Anticipation, and Reflection. For each area the student was invited to answer ten questions on a 5-point Likert-type scale, with 1 corresponding to 'never' and 5 to 'always'. The questions could be formulated in both directions, i.e. either positive or negative. An example of an item in the Organization area is: "sometimes I discover at the last minute that I have not finished my homework"; and an item of Elaboration is "when I study I try to repeat verbatim what I have read in the text." For these two items the score of 'never' was 5 as it reflected a better study method and the score of 'always' was 1, and intermediate responses ('sometimes', 'often', 'very often') were similarly treated. For the area of Prediction, an example is: "I am able to predict if the task will be easy or difficult", for the area of Test Anticipation an example is: "I pay attention to the teachers's questions in order to understand what they want me to know", and for the Reflection area an example is: "If I fail in a task I try to understand the reason why." For these last three items the score of 'never' was 1 and 'always' was 5. The original Questionnaire was greatly reduced and modified in the new adaptation and a pilot administration was necessary to check comprehensibility and the basic psychometric properties of the new version. An administration to 79 students of approximately the same age showed that the adapted and reduced version of the QS was suitable and also had quite good psychometric properties. In particular, Cronbach's alphas for the five areas were 0.77, 0.51, 0.51, 0.67, and 0.58. respectively.

The questionnaire was administered, collectively, to the participants divided in small groups.

Results

Table 3 presents the mean overall scores obtained by the two groups in the five areas of the Study Method Questionnaire. Student's comparisons revealed a significantly better score for the high-coherent students in all five areas, i.e. they were better at organizing their study, developing deeper personal elaboration, evaluating task difficulty, anticipating test-taking and reflecting on their cognitive functioning.

Table 3 also presents the overall evaluations given in the Strategy Questionnaire for the 39 strategies. It can be seen that efficacy ratings had

Table 3. Mean scores obtained by students with a high and a low strategic coherence in the five areas of the Study Method Questionnaire, in the overall summed ratings for efficacy and use in the Strategy Questionnaire, and in an achievement measure (Language mark) (M = mean score, SD = standard deviation).

	High-coherent		Low-coherent		
Variable	*M*	*SD*	*M*	*SD*	$t(39)$
Organization	39.98	4.55	34.43	5.87	5.27 $p < 0.001$
Elaboration	33.47	4.04	30.67	4.13	3.43 $p = 0.001$
Prediction	40.00	4.35	37.10	3.86	3.53 $p = 0.001$
Test anticipation	37.82	4.82	34.96	5.25	2.83 $p = 0.006$
Reflection	34.80	5.19	31.71	4.86	3.07 $p = 0.003$
Efficacy score	189.49	24.48	187.65	21.93	n.s.
Use score	183.61	26.11	145.39	23.65	7.68 $p < 0.001$
Language mark	6.42	0.76	6.08	0.98	1.99 $p = 0.049$

similar values in the two groups, whereas use ratings were significantly higher for the high-coherent group, suggesting that in this case, at least a part of the coherence score was related to the fact that low-coherent students recognized the general value of study strategies but did not use them. The last line of Table 3 shows the mean mark obtained in Language by the two groups of students. It must be noted that the marks were collected on the basis of official school reports at the end of the school year. These marks present a very low range of values, typically between 5 and 7, probably to avoid excessive discrimination between students. For this reason the difference was not particularly great, however it reached the significant level of 0.05.

As we also had the opportunity of administering the Study Method Questionnaire to the subjects who took part in the first (selection) phase of the Study, we made analyses concerning the overall initial sample of 234 students. In fact the inclusion of all subjects could regard the complete distribution of scores in the variables considered and could then test the hypothesis that higher coherence affects study method and a better study method produces higher achievement. We assume that greater strategic coherence reflects higher levels of metacognitive control because control guarantees a better correspondence between metacognitive evaluations and actual action and because self-perception of strategic coherence can determine a higher effortful control. Our analyses tested the hypothesis that better strategic coherence, because of its association with better metacognitive control, should affect the capacity to adopt an effective study method, in particular regarding the components that reflect self-knowledge, self-control and self-evaluation capacities, which, in turn, should lead to greater academic achievement.

Table 4 presents Pearson's correlation between the strategic coherence score, Language mark and the scores in the five areas of the Study Method Questionnaire. Nearly all these variables were significantly correlated, although the correlation values were not particularly high. In particular, the strategic coherence score was correlated with the five Study Method areas, with values ranging from 0.32 (Organization) to 0.15 (Reflection). Furthermore three of these areas (Organization, Elaboration, Prediction) were correlated with the Language mark.

The path analysis, run with the LISREL program, release 8.30 (Jöreskog & Sörbom, 1996), produced the pattern of data illustrated in Fig. 1, with an Adjusted Good Fitness Index (AGFI) of 0.44, $\chi^2 = 232.04$, df = 11, $p < 0.001$. This pattern, corresponding to our predictions, appeared to fit better than the patterns combining strategic coherence, study method and school achievement in different ways. It can be noted that strategic coherence significantly affects all five aspects of study method, whereas only three of the latter, i.e. Organization, Elaboration, and Prediction, affect school achievement.

Table 4. Pearson's correlation values between the strategic coherence score (*COH*), the scores in the five areas of the Study Method Questionnaire (*ORG* = organization; *ELA* = elaboration; *PRE* = Prediction; *TA* = Test Anticipation; *REF* = Reflection) and school achievement (*SA*) obtained with 234 students in the original sample of Study 2.

	COH	*ORG*	*ELA*	*PRE*	*TA*	*REF*	*SA*
COH	–						
ORG	0.32[a]						
ELA	0.27[a]	0.24					
PRE	0.21	0.35[b]	0.28[a]				
TA	0.19	0.50[b]	0.32[a]	0.31[a]			
REF	0.15	0.37[b]	0.30[a]	0.34[b]	0.41[b]		
SA	0.21	0.29[a]	0.21	0.32[a]	0.11	0.09	–

[a] $p < 0.05$
[b] $p < 0.01$.

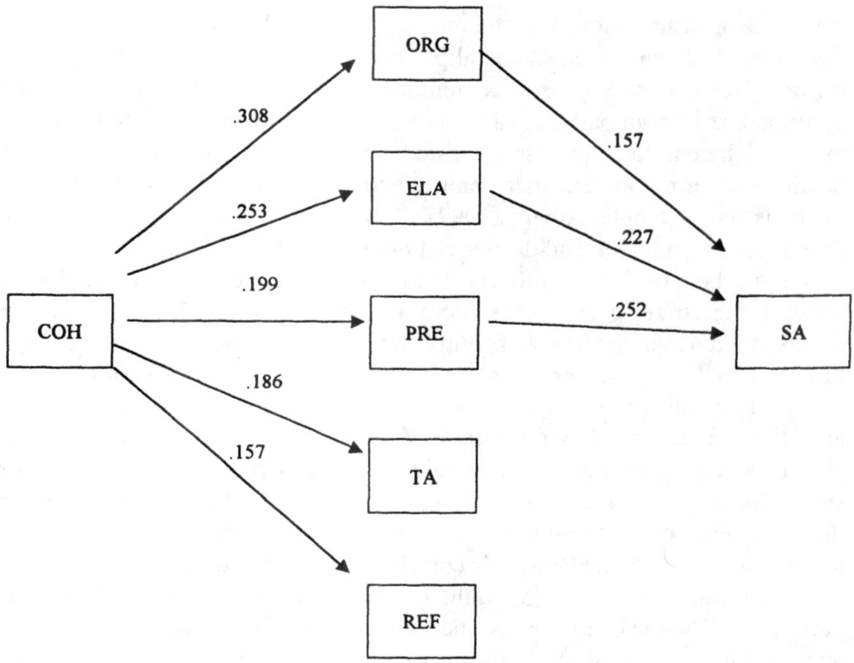

Fig. 1. Causal model illustrating the relationship between strategic coherence (*COH*), study method (*ORG* = organization; *ELA* = elaboration; *PRE* = Prediction; *TA* = Test Anticipation; *REF* = Reflection) and school achievement (*SA*).

Discussion

The results of Study 2 offer a direct insight into the characteristics of students with high or low coherence in their approach to study. Despite its simplicity our measure of strategic coherence was able to individuate two groups of students with very different characteristics. It must be noted that the selection criterion used for this study, i.e. students with a particularly high or low strategic coherence score, led us to find specific groups of subjects only partially overlapping with the groups examined in the previous Study.

Students with low strategic coherence, who offered different ratings for efficacy and use of the same strategies, also had a poorer study methods. Study method was measured on the basis of subjective ratings so that the measures obtained could reflect subjective dissatisfaction, also implied in the strategic coherence measure. However, the measures obtained in the QS Questionnaire have already shown to reflect actual performance in study tasks. Therefore, the lower scores obtained by low-coherent students in the study method measures described, at least in part, actual study behaviors. The less efficient processing by low-coherent students also appeared in the achievement measure as they obtained lower marks. The data confirm and extend previous research on study methods. For example, Albaili (1997), Kleijn et al. (1994) and Turner (1992) found that organization and deeper elaboration improve learning.

It must be noted that, different from Study 1, we found a relationship between the strategic coherence score and strategy use ratings. In fact, low-coherent students gave significantly lower use ratings to the strategies. Furthermore, the two variables were highly correlated (Pearson's $r = 0.71$, $p < 0.001$). Studies 1 and 2 differ because the first was focused on low achievers and the second on low-coherent students. The two variables, although correlated, do not overlap. The strategic coherence scores of the subjects in this study concerned the extreme values only and this had the effect of increasing the probability of selecting students with extreme use ratings (for the original sample of 234 students, Pearson's correlation between strategic coherence and use was in fact lower, i.e. 0.53, although higher than in Study 1, where it was close to 0). The mean absolute value of use ratings in the high-coherent group appeared to be particularly high, for example higher than that given by high achievers in Study 1. However a correlation between use ratings and coherence score appears in line with the hypothesis that an important factor, although not the only one, of low strategic coherence is a low use of study strategies compared with an appropriate estimation of their importance.

Finally, a path analysis of the overall original sample offered evidence sustaining the assumption that strategic coherence has a critical role in study method and effect on achievement. In fact the best pattern of path analysis offered a model showing that strategic coherence may affect the five different aspects of study method considered. The same pattern showed that only three of the latter affected achievement in Language significantly, i.e. ability in Organization, Elaboration and Prediction. In this pattern the aspects of study method regarding Test Anticipation and Reflection were not relevant. This could be due to the fact that students' achievement in Language is less crucially related to their tendency to reflect on the task or anticipate characteristics of the examination. It must be noted that, due to the correlation between strategic coherence scores and composite use ratings we found in this study, it is possible that analyses substituting the two variables should have produced partly similar outcomes.

GENERAL DISCUSSION

School achievement depends on a variety of different factors related to social context, motivation, abilities, personality traits and so on. In this paper we focused on a particular area, i.e. study area, which has also been shown to be critical for achievement (e.g. Cheung & Kwok, 1998; Drew & Watkins, 1998).

In the study context it seems possible to distinguish between different aspects, concerning basic abilities, knowledge, strategies, and attitude towards study. In this paper we focused on a new aspect concerning study, i.e. strategic coherence, and we showed its relevance to educational psychology.

In the present study the role of strategic coherence was demonstrated in three ways. First, examination of the characteristics of a group of low achievers showed that they had particularly low strategic coherence. Secondly, by directly focusing on students with low strategic coherence we were able to show that they also presented other weaknesses in the areas of study method and achievement. Thirdly, in a more general view on a larger sample of students, we found that a model with a good fit was based on the assumption that strategic coherence can affect study method which can, in turn, affect school achievement.

Strategic coherence may appear to be a very specific and narrow concept, but it seems to grasp some critical elements of the general attitude toward study. Evidence collected in the two studies confirms the importance of this aspect and, in our view, offers a partially new perspective to variables related to poor study methods and poor achievement. This perspective extends the number of context-related variables to the student's cognitive, personality and motivational characteristics, and specifically to the student's study method, which altogether contribute to academic achievement.

In fact, it has been hypothesized (e.g. Loranger, 1994) that students with low achievement could be unaware of the best study strategies and the fact that their personal study method is inadequate. This position of "blissful ignorance" seems to be contradicted by our data showing that late adolescents who are low achievers present a higher discrepancy between efficacy and use ratings, i.e. they are aware of the fact that they do not use study strategies to the degree they think necessary. Obviously our data only concern a limited age range (18–19 years) and it is possible that younger low achievers are in some respects less aware of good strategies. Furthermore our data do not offer clear information about the causal direction of the variables we studied. Some evidence collected by De Beni and Moè (1997), who found that an improvement in strategic coherence affects learning positively, suggests that strategic coherence may have a causal role in school achievement.

The strategic coherence measure we collected can be considered a specific aspect of the self-perception measures studied by the literature with reference to the distance between ideal and actual selves. Actually it has been shown that a better social adjustment is associated with a lower discrepancy between the two selves (e.g. Higgins, 1987). Obviously a specific discrepancy in the study context should not be generalized to other contexts and may be strictly connected to students' experiences, in particular failure, success, stimulation and opportunities in the specific area of academic studies. Some clinical observations (De Beni & Moè, 1997) suggest that a low strategic coherence may be related to other emotional and motivational difficulties. In particular, low coherence could be due to a higher motivation to learn and succeed in school, especially when related to a greater use of strategies. However, Study 1 shows that this is not always true, as low achievers had a lower strategic coherence but not a significantly lower strategy use. In this case, other factors must be found to explain why low achievers are less coherent, probably related to a well organized and coherent metacognitive theory of study and cognitive skills, like planning and the ability to control behavior to make it more coherent with that theory.

Further evidence will be necessary in order to clarify better the issues raised in this research. For example, in the present research we mainly used questionnaires and subjective ratings and these ratings are possibly affected by a series of biases. Future research will try to find more objective measures of study strategies and effectiveness. Furthermore, it will be necessary to clarify the extent to which coherence measure adds information to that given by strategy use ratings. In fact, although low strategic coherence was sometimes due to a use of strategies which was higher than the rated efficacy, for most students, especially in study 2, strategic coherence was related to a higher rated use of strategies. This result does not contradict the assumption that strategic coherence is critical as it is related to a discrepancy between recognition of the importance of study strategies and an effective use of them. It suggests, however, that a simple strategy use measure could suffice in some cases.

From a practical educational point of view, we think that our data offer a partially new perspective to interventions with students in difficulty. In fact, it is a common idea (e.g. Wade, Trathen & Schraw, 1990) that low achievers are unaware of good strategies or do not recognize their importance. This becomes the basis of methods for improving study skills which often focus on these aspects. Our data suggest that this approach may not be useful, or may even be confusing, as low achievers may already have this kind of awareness. It seems important to develop a more general student awareness of study methods and help them recognize their contradictions and the reasons why known strategies are not systematically used, even if their importance is recognized.

ACKNOWLEDGMENTS

The research was supported by a MURST Grant to the second author. We are very grateful to Lucia Cacciò, Manuela Provenzano and Claudia Zamperlin for helping us in collecting data.

REFERENCES

Albaili, M. A. (1997). Differences among low-, average- and high-achieving college students on learning and study strategies. *Educational Psychology, 17*(1–2), 171–177.

Bernardi, L., & Valente, N (1998). Il calcolo degli abbandoni. L'indicatore di produttività [The abandonment estimate. The productivity indicator]. *ProgettoBo, 8*, 4–7.

Beishiuzen, J. J., & Stoutjesdijk, E. T. (1999). Study strategies in a computer assisted study environment. *Learning and Instruction, 9*, 281–301.

Cheung, C. K., & Kwok, S. T. (1998). Activities and academic achievement among college students. *Journal of Genetic Psychology, 159*(2), 147–162.

Cornoldi, C. (1995). *Metacognizione e apprendimento* [Metacognition and learning]. Bologna: Il Mulino.

Cornoldi, C., De Beni, R., & MT Group (1993). *Imparare a studiare* [Learning to study]. Trento: Erickson.

De Beni, R., & Moè, A. (1997). Difficoltà di studio. Un intervento metacognitivo con studenti universitari [Study difficulties. A metacognitive training with university students]. *Psicologia Clinica dello Sviluppo, 1*(3), 433–440.

Drew, P. Y., & Watkins, D. (1998). Affective variables, learning approaches and academic achievement: A causal modelling investigation with Hong Kong tertiary students. *British Journal of Educational Psychology, 68*(2), 173–188.

Gadzella, B. M. (1995). Differences in processing information among psychology course grade groups. *Psychological Reports, 77*(3, pt. 2), 1312–1314.

Herrmann, D., Raybeck, D., & Gutman, D. (1993). Improving student memory. Seattle, WA: Hogrefe & Huber.

Hettich, P. I. (1998). *Learning skills for college and career* (2nd ed.). Pacific Grove, CA: Brooks/Cole.

Higgins, E. T. (1987). Self-discrepancy: A theory relating self and affect. *Psychological Review, 94*, 319–340.

Huet, N., & Mariné, C. (1996). Assessment of metamemory knowledge and use of strategy. *Psychological Reports, 79*, 1203–1206.

Jöreskog, K., & Sörbom, D. (1996). *LISREL 8: User's reference guide*. Chicago, IL: Scientific Software International.

Kleijn, W. C., Van der Ploeg, H. M., & Topman, R. M. (1994). Cognition, study habits, test anxiety, and academic performance. *Psychological Reports, 75*(3, pt 1), 1219–1226.

Leondari, A., Syngollitou, E., & Kiosseoglou, G. (1998). Academic achievement, motivation and future selves. *Educational Studies, 24*(2), 153–163.

Loranger, A. L. (1994). The study strategies of successful and unsuccessful high school students. *Journal of Reading Behavior*, *26*(4), 347–360.

Markus, H., & Nurius, P. (1986). Possible selves. *American Psychologist*, *41*, 954–969.

McCutchen, D., Francis, M., & Kerr, S. (1997). Revising for meaning: Effects of knowledge and strategy. *Journal of Educational Psychology*, *89*(4), 667–676.

Nolen, S. (1988). Reasons for studying: Motivational orientations and study strategies. *Cognition and Instruction*, *5*(4), 269–287.

Oyserman, D., Gant, L., & Ager, J. (1995). A socially contextualized model of African American identity: Possible selves and school persistence. *Journal of Personality and Social Psychology*, *69*(6), 1216–1232.

Pressley, M., Van Etten, S., Yokoi, L., Freebern, G., & Van Meter, P. (1998). The metacognition of college studentship: A grounded theory approach. In: D. J. Hacker, J. Dunlosky, J., & A. C. Graesser (Eds), *Metacognition in Educational Theory and Practice. The Educational Psychology Series* (pp. 347–363). Mahwah, NJ: LEA.

Pressley, M., Yokoi, L., Van Etten, S., & Freebern, G. (1997). Some of the reasons why preparing for exams is so hard: What can be done to make it easier? *Educational Psychology Review*, *9*(1), 1–38.

Stoynoff, S. (1996). Self-regulated learning strategies of international students: A study of high- and low-achievers. *College Student Journal*, *30*(3), 329–336.

Stoynoff, S. (1997). Factors associated with international students' academic achievement. *Journal of Instructional Psychology*, *24*(1), 56–68.

Terreni, A., & Campiotti, E. (1999). Deficit metacognitivi nell'attività di studio di adolescenti con sintomatologia depressiva [Metacognitive study deficits in depressed adolescents]. *Psicologia Clinica dello Sviluppo*, *3*(1), 61–80.

Turner, G. Y. (1992). College students' self-awareness of study behaviors. *College Student Journal*, *26*(1), 129–134.

Van Etten, S., Freebern, G., & Pressley, M. (1997). College students' beliefs about exam preparation. *Contemporary Educational Psychology*, *22*, 192–212.

Van Etten, S., Pressley, M., Freebern, G., & Eschevarria, M. (1998). An interview study of college freshmens' beliefs about their academic motivation. *European Journal of Psychology of Education*, *13*(1), 105–130.

Wade, S. E., Trathen, W., & Schraw, G. (1990). An analysis of spontaneous study strategies. *Reading Research Quarterly*, *25*(2), 147–166.

Weinstein, C. E., & Hume L. M. (1998). *Study strategies for lifelong learning*. Washington, D.C.: APA.

Wilding, J., & Valentine, E. (1992). Factors predicting success and failure in the first-year examinations of medical and dental courses. *Applied Cognitive Psychology*, *6*, 247–261.

Wolters, C. A. (1998). Self-regulated learning and college students' regulation of motivation. *Journal of Educational Psychology*, *90*(2), 224–235.

Wood, E., Motz, M., & Willoughby, T. (1998). Examining students' retrospective memories of strategy development. *Journal of Educational Psychology*, *90*(4), 698–704.

APPENDIX 1

Strategies presented in the Strategy Questionnaire listed according to first phase presentation order (efficacy ratings). Presentation order in the second phase (use ratings) is specified by the number associated to each strategy. The strategies considered as particularly effective or ineffective in Study 1 are indicated by one (*) or two (**) asterisks respectively.

10. Think about concepts or data already known about the topic.
31. Choose a different way of studying according to the kind of text and goals.
23. Plan daily and weekly study activities, time to spend and length of study sessions.
19. Survey the book or chapter before reading.
33. Read the text aloud. (**)
3. Underline during reading. (**)
39. Underline using different colored pens or pencils.
34. Make notes while studying. (*)
17. Listen to the radio or music while studying. (**)
6. Take care of personal welfare (food, sleep, times). (**)
14. Study in common places to concentrate better and take examples from others. (**)
27. While reading try to foresee the content of the following paragraphs or chapters.
30. While studying summarize by writing in own words.
36. While studying repeat the main points in own words.
21. Try to personalize the main points by asking self about their personal relevance.
12. If possible, find concrete examples of the main points.
11. Summarize the chapter by writing after studying it.
1. Make schema and diagrams. (*)
32. Write guiding concepts next to the text. (*)
8. Try to memorize main points using mnemonic techniques (associations, rhymes, phonetic method, bizarre links, and so on).
9. Try to memorize names, numerical or technical data using mnemonic techniques (associations, rhymes, phonetic method, bizarre links, and so on). (*)
15. Check for comprehension. (*)
5. Immediately change the method of studying if unsure of having understood.

29. In case of difficulty in comprehension, do not care about the concepts not understood or stop studying.
18. Be careful and use strategies to avoid distractions.
28. Find information about possible exam questions.
24. Imagine possible exam questions.
7. Try to repeat what has already been read before finishing study.
26. Repeat after having finished study.
35. Review material after an extended period of time. (*)
4. Study with a friend. (**)
22. Review with a friend. (**)
2. Interpolate the study of other issues, in the study of a chapter. (**)
37. Underline the points remembered less well and review them later. (*)
13. Try to integrate with other texts or supports.
25. Select a particular interesting idea and extend it further than what is written in the text.
16. Ask oneself what the most interesting and/or most critical points of the chapter are.
20. Decide not to study some parts of the chapter because less interesting or important.
38. Simulate characteristics of the examination in the imagination, to be prepared to manage anxiety and unforeseen situations. (*)

SUBJECT INDEX

www.ingramcontent.com/pod-product-compliance
Lightning Source LLC
LaVergne TN
LVHW010650300926
849161LV00005B/1087